Science Versus Textbook Evolution

The Mechanisms of Evolution

Mace Baker, Ph.D.

New Century Books
Lakeland, Florida

About the Author

Mace Baker has taught young people for over 20 years in the public school system. He earned his B.A. from Biola University, and his Ph.D. from Pacific University in Missouri. He has been engaged in Creation Research for more than 20 years and has been conducting seminars in America, Canada, Mexico, England, Western and Eastern Europe, Russia, the Philippines, India, and Central America.

Cover design by Darryl Hodson

Acknowledgments:

Many thanks to Ed and Charlene Rosa and to Darryl Hodson for their very helpful preliminary proof-reading of this volume, and I am especially indebted to Melissa Pagnucco whose excellent secretarial work and editing has helped bring this volume to completion. This has included assistance in typing, proof-reading, and formatting. Her carefulness and attention to detail has been accompanied by a wonderful attitude, which has made each work session a positive experience.

Printed and bound in the United States of America

For information on Dr. Mace Baker's lectures and seminars, the author may be contacted at macebaker@myyellowstone.net

ISBN 0-9647283-5-4

Introduction

This book is written for the purpose of making it possible for parents of students in the public schools, and for the public at large, to be aware of the evolutionary propaganda, which is present in all of the state-adopted high school biology texts. It is also written for the purpose of showing the tremendous amount of scientific evidence that is antagonistic to all of this evolutionary propaganda. It will be noted that the objective evidence in the world of nature is clearly antagonistic to the evolution model, but not to the Creation model, as presented in the book of Genesis.

It is important for me to be very clear about one aspect of this book. The vast majority of quotes, which are antagonistic to the theory of evolution, are actually quotes from evolutionists. There is only one quote from a known Christian, Dr. Stephen Meyer of the Discovery Institute, and probably on only one or two occasions, do I quote from an individual, such as Michael Denton, who is technically not an evolutionist. He and his kind have been associated with the title "non-creationists." This means that he is neither a Christian nor a creationist, yet he has written a book showing the flaws in the theory of evolution.

It is also important that I emphasize this point very forthrightly--I am not by any means implying that the evolutionists, which I quote, have changed their mind about the value of evolution and have become sympathetic to the idea of creation. Each one that I quote, as best as I can ascertain, is indeed a loyal evolutionist and is totally committed to the theory of evolution. They are simply stating accurate facts of nature with respect to the particular issue at hand, and though the information they relate is unsupportive of this theory, they themselves would readily confess that they firmly believe in the consensus--that is, in the theory of evolution.

Hopefully the objective readers will find after reading this volume, as well as volume two of *Science Vs. Textbook Evolution*, that it is not unreasonable to conclude that the theory of evolution is upheld primarily by this consensus and not by the facts of nature.

TABLE OF CONTENTS

ORIGIN OF LIFE

Introduction

Generally speaking, the modern high school biology textbooks acknowledge that there are three ways life could have begun on the earth: extraterrestrial origin, origin by divine creation and spontaneous generation from non-living matter. The first two are immediately dismissed by the textbook authors as not being able to be tested and are therefore out of the realm of science. The implication is, of course, that origin from non-living matter can and indeed has been tested and, in the course of the text, students are informed that the "scientific" tests have demonstrated that life could certainly have arisen from non-living matter by totally naturalistic processes.

As George B. Johnson implies in the Holt, Rinehart, & Winston text, *Biology Visualizing Life,* a Supernatural Creator is not needed for the explanation of how life originated on the primitive earth,

> "The basic building blocks of life can assemble spontaneously, requiring no form more mysterious than simple chemistry."[1]

The concept of the origin of life from non-living matter began more than a hundred years ago when many people in the western hemisphere began to adopt, as a preferential religio-philosophy, the idea of materialism. This concept espouses the belief that matter constitutes the whole of the universe; that is to say, there is no Creator God, nor any real world of spiritual life.

> "Evolutionary theory is, in the minds of many biologists, the foundation on which all biological science is built."[2] (Miller & Levine, *Biology*, Prentice Hall)

By and by, this philosophy began to gain more and more adherents, many of whom originally (though often quite superficially) had held to the idea of a Divine Creator. After having accepted the idea of materialism, it then became important to explain in some "scientific" manner how life, apart from Divine Creation, could have come about.

Textbook Evolution

One of the first and most important contributors to this theory was the Russian scientist A.I. Oparin. The students read of Oparin's theory in the Glencoe text, *Biology The Dynamics of Life*,

> "In the 1930's, a Russian scientist, Alexander Oparin, proposed a widely accepted hypothesis that life began in the early oceans. He suggested that energy from the sun and from lightning triggered chemical reactions to produce simple organic compounds from the substances present. Oparin envisioned many chemical reactions occurring in the atmosphere and the products raining down into the oceans to form what is often called a primordial soup."[3] (Macmillan/ McGraw Hill)

As reported in another state-adopted text, a further assumption, in addition to the primordial soup postulation, had to be made with regard to the early Earth's atmosphere that it contained no oxygen,

> "Oparin, Harold Urey of the University of Chicago, and other investigators of the solar system, proposed that Earth's early atmosphere lacked oxygen. They hypothesized that the early atmosphere was instead rich in nitrogen (N_2) and hydrogen containing gases such as hydrogen (H_2), water vapor (H_20), methane (CH_4), and ammonia (NH_3)."[4] (George B. Johnson & Peter H. Raven, *Biology: Principles and Explorations*, Holt, Rinehart and Winston, Inc.)

Albert Towle, in the *Modern Biology* text, comments on the problem of having oxygen in the atmosphere of the early Earth,

> "Though this substance is essential to most forms of life today, it would have destroyed organic compounds in the hot atmosphere of the young Earth by the chemical process called oxidation"[5] (Albert Towle, *Modern Biology*, Holt, Rinehart & Winston)

> "Because Earth's atmosphere lacked oxygen, scientists have proposed that these organisms were most likely anaerobic."[6] (Alton Biggs et al., *Biology: The Dynamics of Life*, Glencoe/McGraw-Hill)

So the naturalistic theories that arose argued that life began on the Earth at a time when the Earth's atmosphere contained a variety of gases such as ammonia and methane but was devoid of oxygen. The earliest life forms evolved out of a vast ocean of "primordial soup" which had spontaneously evolved vast amounts of the varied organic molecules essential to life.

> "Oparin envisioned these molecules forming spontaneously in chemical reactions activated by energy from solar radiation, volcanic eruptions, and lightning. Oparin thought that over millions of years, these molecules had gradually come together to form living matter."[7] (George B. Johnson & Peter H. Raven, *Biology: Principles and Explorations*, Holt, Rinehart and Winston, Inc.)

The Addison-Wesley text states that,

> "...studies suggest that Earth's atmosphere 4 billion years ago probably consisted of gases such as ammonia (NH3), methane (CH4), carbon dioxide (C)2), and nitrogen (N2)."[8] (Bernice Essenfeld, Carol Gontang, & Randy Moore, *Biology*, Addison-Wesley)

With regard to oxygen, they further state,

> "Evidence indicates that there was little or no oxygen (02) in the atmosphere when life appeared"[9] (*Ibid.*)

Once again, the texts are telling the students that "evidence" shows that the atmosphere of the early earth consisted of much ammonia and methane gases but no oxygen. We will soon find that just the opposite is true.

Next, it became necessary to develop an experiment which would demonstrate the feasibility of Oparin's life-from-non-life idea. In George B. Johnson's *Biology: Visualizing Life*, we read,

> "What chemical reactions could occur in this atmosphere? To answer this question, in 1953 Stanley Miller and Harold Urey re-created the conditions of the early Earth inside two connected

flasks, as illustrated in Figure 10.2. The atmosphere Miller and Urey introduced into the flasks contained simple molecules that were probably found in the early Earth's atmosphere: hydrogen, carbon dioxide, methane, water vapor, ammonia, and carbon monoxide. They heated the mixture and zapped it with electrical sparks to simulate lightning. Within days, a dark, smelly mixture developed. When this mixture was analyzed, Miller and Urey found that many complex molecules had formed, including some amino acids, the building blocks of proteins."[10] (Holt, Rinehart, & Winston)

"Figure 17-8 Miller and Urey produced amino acids, which are needed to make proteins, by passing sparks through a mixture of hydrogen, methane, ammonia, and water. This and other experiments suggested how simple compounds found on the early Earth could have combined to form the organic compounds needed for life."[11] (Kenneth R. Miller & Joseph Levine, *Biology*, Pearson/Prentice Hall)

Miller and Levine report,

"Figure 16-8. An experiment performed by Stanley Miller (top) and Harold Urey first demonstrated how organic matter may have formed in Earth's primitive atmosphere. By re-creating the early atmosphere (ammonia, water, hydrogen, and methane) and passing an electric spark (lightning) through the mixture, Miller and Urey proved that organic matter such as amino acids could have formed spontaneously...."[12] (Kenneth R. Miller and Joseph Levine, *Biology*, Prentice Hall)

If there was no Oxygen, then there was no Ozone Layer. The presence of Oxygen poses a large problem for evolutionary theorists. Consequently, some textbooks state emphatically that there was no ozone layer.

"Some energy sources present on the surface of early Earth are absent or reduced today. For example, the atmosphere lacked an ozone layer such as the one that now surrounds Earth and shields us from the sun's harmful ultraviolet radiation."[13] (*Ibid.*)

"There were energy sources on the surface of the early Earth that are absent or reduced today. For example the primitive atmosphere had no ozone layer. Without an ozone layer, enormous amounts of energy, in the form of ultraviolet radiation, reached the surface of the Earth."[14] (Bernice Essenfeld, Carol Gontang, & Randy Moore, *Biology*, Addison-Wesley)

What is amazing is that right after the sentence stating that the early earth had no ozone layer, they also state,

> "Without an ozone layer, enormous amounts of energy in the form of ultraviolet radiation reached Earth's surface."[15] (*Ibid.*)

As a result of all these aforementioned theories and the Stanley Miller experiment, the *Biology: Visualizing Life* text, as quoted earlier, tells the students that,

> "The basic building blocks of life can assemble spontaneously, requiring no form more mysterious than simple chemistry."[16]

In other words, according to these high school textbooks, one does not have to appeal to a Supernatural Creator for the development of the smallest building blocks of life, they can just come together by chance processes.

Science Versus

In the years following the Urey-Miller experiment, a considerable amount of research has been conducted. The results of this research has greatly undermined Oparin's theory and the Urey-Miller experiment. Among others, *Time* magazine has alerted the general public to the fact that the aforementioned theories are not floating smoothly in the scientific community,

> "Now this textbook picture of how life originated, so familiar to college students just a generation ago, is under serious attack."[17]

In general, it has been demonstrated that Miller's experiment has three insurmountable problems. It had the wrong starting ingredients, the wrong conditions and obtained the wrong results.

Furthermore, it is extremely important to note that in all these experiments, very elaborate lab equipment is needed. Precisely measured amounts of chemicals are used, the environments are devoid of contaminant chemicals, prolonged sources of energy "unlike anything suggested in nature" are also required. Further, a highly educated and an intelligent being is essential to carefully control the entire procedure. Consequently, to imply that chance and time is all that is needed is a most unreasonable assumption.

Wrong Starting Ingredients

Miller's experiment (which embraced Oparin's theory) had the early atmosphere containing ammonia and methane, but no oxygen. Research shows that just the opposite was true. If the atmosphere of the early Earth contained methane, it should now be found in association with ancient sedimentary clays; however, it is not to be found. The ammonia postulated would have easily dissolved in the early Earth's oceans (750 gallons of ammonia can be dissolved in one gallon of water). For ammonia to have been present in an adequate amount, there would have had to have been a giant container of ammonia circling the Earth, in order to have insured that there would be enough left over, after the absorption process, to fill the atmosphere.

The *Time* magazine article also informs us that,

> "New insights into planetary information have made it increasingly doubtful that clouds of methane and ammonia ever dominated the atmosphere of the primitive Earth... if scientists have, by and large, tossed out the old ideas, they have not yet reached the consensus on a new."[18]

At a much earlier date the geochemist, P. Abelson's, question and answer indicated that the non-oxygen atmosphere idea was largely imagination,

> "What is the evidence for a primitive methane-ammonia atmosphere on Earth? The answer is that there is no evidence for it, but much against it."[19]

Furthermore, Abelson, shows that the methane/ammonia scenario would have been of no value,

"The hypothesis of an early methane-ammonia atmosphere is found to be without solid foundation and indeed is contradicted."[20]

The idea of the early earth containing no, or precious little, oxygen has been presented by the textbooks, as cited above, and has been effective in convincing many people that this is a true fact of science. For instance, consider this sentence from the book, *The Invention of Air: A Story of Science, Faith, Revolution, and the Birth of America*, by Steven Johnson, "We know that the gas that Priestly was observing was dioxygen, otherwise known as free oxygen or o2, a molecule formed by the union of two oxygen atoms. While oxygen is the third most common element in the universe we know that free oxygen was exceedingly rare in the earth's initial atmosphere."[21]

Actually, much recent research demonstrates that the early Earth's atmosphere contained oxygen. In his address to a symposium on the evolution of the earth's early atmosphere, the British geologist, C. F. Davidson's remarks included,

"These observations based on the stratigraphical record contradict the belief that a primeval reducing atmosphere persisted for much of Pre-Cambrian time."[22]

Molecular biologist, Michael Denton, also reports,

"...rocks of great antiquity have been examined over the past two decades and in none of them has any trace of abiotically produced organic compounds been found...."[23]

Further, as shown by Erich Dimroth and Michael M. Kimberly writing in an article for *Canadian Journal of Earth Sciences*, the rocks show that oxygen has been with us from the beginning,

"The sedimentary distribution of carbon, sulfur, uranium and ferric and ferrous iron depend greatly upon ambient pressure and should reflect any major change in the proportion of oxygen in the atmosphere or hydrosphere. The similar distributions of these elements in sedimentary rocks of all ages are here interpreted to indicate the existence of a Precambrian atmosphere containing much oxygen...we find no evidence...that an oxygen free atmosphere has existed at any time during the span of geological history recorded in well preserved sedimentary rocks."[24]

Clemney, Badham, and others further report that,

> "The existence of early red beds, sea and groundwater sulphate, oxidized terrestrial and sea-floor weathering crusts, and the distribution of ferric iron in sedimentary rocks are geological observations and inferences compatible with the biological and planetary predictions. It is suggested that from the time of the earliest dated rocks at 3.7 b.y. ago, Earth had an oxygenic atmosphere."[25]

Describing the works of NASA's atmospheric scientist, Dr. Joel Levine, Patrick Huyghe's writes,

> "Another area in the traditional theory of the origin of life that has been thrown into question by Levine and his colleagues is the level of oxygen in the prebiological atmosphere...it now appears that the early atmosphere may have contained a million times more oxygen than anyone had believed before...The early geological record appears to support this conclusion. Geologists have long suspected that the oxygen level in the early atmosphere had to be much greater than previously calculated. Their analysis of the oldest known rocks, estimated to be more than 3.5 billion years old, found oxidized iron and uranium in amounts that called for the atmospheric oxygen levels to be from a hundred to a billion times greater than accepted."[26]

Huyghe reports that the early atmosphere may have contained a million, even a billion, times more oxygen than had been predicted. The *Bulletin of the American Meteorological Society* has also reported that early earth oxygen levels may have been a billion times greater than expected,

> "In the case of our calculated oxygen levels, one bit of evidence from the early geological record supports our conclusion. It was puzzling, but the geologists know from their analyses of the oldest known rocks that the oxygen level of the early atmosphere had to be much higher than previously calculated. Analyses of these rocks estimated to be more than 3.5 billion years old, found oxidized iron

in amounts that called for atmospheric oxygen levels to be at least 110 times greater and perhaps up to one billion times greater than otherwise accepted."[27]

Again, Clemmey and Badham report,

"Geologic evidence often presented in favor of an early anoxic atmosphere is both contentious and ambiguous...."[28]

"Recent biological and interplanetary studies seem to favor an early oxidized atmosphere rich in CO_2 and possibly containing free molecular oxygen...."[29]

A report in the *New Scientist* states,

"Although biologists concerned with the origin of life often quote an early atmosphere consisting of reduced gases, this stems as much from ignorance of recent advances as from active opposition to them....The time has come, it seems, to accept as the new orthodoxy the idea of early oxidized atmospheres on all three terrestrial planets, and the biological primer which still tells of life on Earth starting out from a methane/ammonia atmosphere energized by electric storms and solar ultraviolet need to be rewritten."[30]

For many scientists,

"...the concept of a reducing primitive atmosphere has been abandoned."[31]

Sidney Fox and Klaus Dose also concluded that the idea of a reducing atmosphere did,

"...not seem to be geologically realistic because evidence indicates that...most of the free hydrogen probably had disappeared into outer space and what was left of methane and ammonia was oxidized."[32]

It is nice that at least one textbook admits to the fact that without oxygen, ultraviolet radiation would reach the Earth. This poses an insurmountable problem to the theory that the early Earth contained no oxygen.

> "Without oxygen there would be no ozone layer in the upper atmosphere which today protects the Earth's surface from a lethal dose of ultraviolet radiation. In any oxygen-free scenario, the ultraviolet flux reaching the Earth's surface might be more than sufficient to break down organic compounds as quickly as they were produced. Significantly, the absence of organic compounds in the Martian soil has been widely attributed to just such a strong ultraviolet flux which today continuously bombards the planet's surface. What we have then is a sort of 'Catch 22' situation. If we have oxygen we have no organic compounds, but if we don't have oxygen we have none either."[33]

In order to make their non-life to life scenario work, they must first assume that there was no oxygen on the early Earth. This, then, forces the promoters of this theory to have to make a secondary assumption that there was no ozone or, a tertiary assumption that the ultra-violet light was not as strong as it is now.

The late Carl Sagan has also previously acknowledged this problem and has, more or less wistfully suggested, that the ultraviolet radiation in the past could not have been at its present rate.

> "Unacceptable high mutation rates will, of course occur at much lower u.v. doses, and even if we imagine primitive organisms having much less stringent requirements on the fidelity of replication than do contemporary organisms, we must require very substantial u.v. attenuation for the early evolution of life to have occurred..."[34]

Finally, as Francis Hitching, a British author and journalist, admits, with or without oxygen, the 'basic building blocks of life' could not get started. Thus the transition from non-life to life, on an evolutionary basis is essentially a matter of imagination,

> "So with oxygen in the air, the first amino acid would never have got started; without oxygen, it would have been wiped out by cosmic rays."[35]

In addition, he points out that,

"Imaginative and elaborate solutions have been written to this conundrum. Perhaps the amino acid was formed at the edge of a volcano, and then sank into a lake where it dropped the few meters below the surface necessary to protect it from radiation; perhaps the Earth's waters were covered by a layer of tar-like chemicals which stopped ultraviolet light; perhaps the amino acid was protectively dehydrated or 'frozen' in some way on dry rock or clay, waiting for an improvement in the atmosphere. For every suggestion, there is a seemingly insuperable objection: beneath the surface of the water there would not be enough energy to activate further chemical reactions; water in any case inhibits the growth of more complex molecules; unlike conditions in laboratory experiments, the amino acids and their constituents could not be kept pure and isolated. In other words, the theoretical chances of getting through even this first and relatively easy stage in the evolution of life are forbidding."[36]

Further, he also points out that efforts to get around this problem have been unsuccessful. The evidence has become quite clear that the textbook evolution scenario of an early Earth with no oxygen is essentially anti-science. In addition to this, the "primordial or prebiotic soup" idea is equally devoid of evidence. Earlier we quoted Oparin who,

"... suggested that energy from the sun and from lightning, triggered chemical reactions to produce simple organic compounds from the substances present. Oparin envisioned many chemical reactions occurring in the atmosphere and the products raining down into the oceans to form what is often called a primordial soup."[37]

Past Professor of Biology, Harold Blum, acknowledges that,

"It is commonly assumed today that life arose in the oceans... But even if this soup contained a goodly concentration of amino acids, the chances of their forming spontaneously into long chains would seem remote. Other things being equal, a dilute hot soup would seem a most unlikely place for the first polypeptides to appear... The spontaneous formation of a polypeptide of the size of the smallest known proteins seems beyond all probability."[38]

Further, the *Encyclopedia of Evolution*, states:

"Decades of persistent failure to create life by the 'spark in the soup' method (or to find such productions in nature) have caused some researchers to seek other approaches to the great enigma."[39]

Denton emphasizes the enigma,

"Considering the way the prebiotic soup is referred to in so many discussions of the origin of life as an already established reality, it comes as something of a shock to realize that there is absolutely no positive evidence for its existence."[40]

Further, Professor Noam Lahav of the Hebrew University in Jerusalem reiterates the same observation,

"Under slightly reducing conditions, the Miller-Urey action does not produce amino acids, nor does it produce the chemicals that may serve as the predecessors of other important biopolymer building blocks. Thus, by challenging the assumption of a reducing atmosphere, we challenge the very existence of the "prebiotic soup", with its richness of biologically important organic compounds. Moreover, so far, no geochemical evidence for the existence of a prebiotic soup has been published. Indeed, a number of scientists have challenged the prebiotic soup concept, noting that even if it existed, the concentration of organic building blocks in it would have been too small to be meaningful for prebiotic evolution."[41]

Note the assignment that the students are asked to do in the Addison-Wesley 1994 text,

"The chemical ingredients for the first organic molcecules are sometimes referred to as the 'primordial soup.' Write a recipe for primordial soup. Give all necessary ingredients and instructions. Explain your analogy."[42]

Consider the futility of such an assignment being given to a high school student in light of the statements offered above by professional researchers, as well as the following by world-renowned scientist, Freeman Dyson,

"We are still at the very beginning of the quest for understanding of the origin of life. We do not yet have even a rough picture of the nature of the obstacles that prebiotic evolution has had to overcome. We do not have a well defined set of criteria by which to judge whether any given theory of the origin of life is adequate."[43]

Astronomer Fred Hoyle is quite dogmatic in his assessment, "There is not a shred of objective evidence to support the hypothesis that life began in an organic soup here on the Earth."[44]

Chemists, Jeffrey Bada and Antonio Lazcano, state,

"But is the "prebiotic soup" theory a reasonable explanation for the emergence of life? Contemporary geoscientists tend to doubt that the primitive atmosphere had the highly reducing composition used by Miller in 1953."[45]

We could have no better summary for this matter than that provided by A.G. Fisher of Princeton University,

"Both the origin of life and the origin of major groups of animals remain unknown."[46]

Wrong Conditions

In all Origin of Life experiments in which amino acids have been produced, the researchers had to develop a means of quickly getting the end product away from the artificial energy source. If the newly developed amino acids would have been left for any significant amount of time in the same environment in which they had been produced (by the electrical sparks), they would have been immediately destroyed by the same source.

In Miller's famous experiment, the end products were quickly trapped out and isolated so that they would not be destroyed. If this procedure had not been developed, Miller's experiment would have produced no noteworthy results.

Actually, due to the presence of solar radiation on the early Earth, any such compounds that would have been formed by lightning, would have been destroyed at least 10,000 times faster by means of solar radiation. As reported by the *Bulletin of the American Meteorological Society,*

> "Ultraviolet radiation on the earth from the young sun may have been up to 100,000 times greater than today."[47]

On the early earth there was no one around to trap out these compounds and to hide them from the ultraviolet rays, cosmic rays and electrical discharges.

Miller himself admits,

> "Most of the photochemical reactions at these low wavelengths would have taken place in the upper atmosphere. The compounds so formed would have absorbed at longer wavelengths and therefore might have been decomposed by this ultraviolet light before reaching the oceans."[48]

D.E. Hull's remarks in an article for *Nature* also illustrate the obvious evidence against the theorized conditions of Miller's experiment,

> "They have merely used the well known principle of increasing the yield of a reaction by selectively removing the product from the reaction mixture. The mere fact that a chemist can carry out an organic synthesis in the laboratory does not prove that the same synthesis will occur in the atmosphere or open sea without the chemist"[49]

It is theorized that such compounds would have been formed in the atmosphere. Of course, they would have also been destroyed in the atmosphere. Yet, since life is here and the idea of special creation is ruled out, the evolutionary theorists must assume that somehow some of the product must have made it into the oceans. There, in the primeval oceans they supposedly grew into self replicating cells. However it is now understood that thermodynamic and hydroelectric reactions in the water would have destroyed the compounds long before they could have evolved into cells and organisms.

Let's consider for a moment the basic components for a single celled organism to live. A well-versed high school biology student should be aware of these basic requirements: a cell barrier or membrane to regulate and control what enters and leaves the cell in order to maintain the cell's integrity and chemical balance, some energy harnessing mechanism or metabolic motor whereby the cell receives its energy supply to perform all the essential work or functions of the cell, the complex series of enzyme sequences required to trigger and regulate life's biochemical reactions and a means of reproducing, which requires the complex sequence of DNA or RNA which houses the entire genetic make-up of the organism and all of its parts and chemistry. One can see from these few basic requirements that this is a far cry from the few elemental parts the laboratory experiments have demonstrated.

Michael Denton, puts this in a picturesque way,

"To grasp the reality of life as it has been revealed by molecular biology, we must magnify a cell a thousand million times until it is twenty kilometers in diameter and resembles a giant airship design. On the surface of the cell we would see millions of openings, like the portholes of a vast space ship, opening and closing to allow a continual stream of materials to flow in and out. If we were to enter one of these openings we would find ourselves in a world of supreme technology and bewildering complexity....Is it really credible that random process would have constructed a reality, the smallest element of which--a functional protein or gene--is complex beyond our own creative capacities, a reality which is the very antithesis of chance, which excels in every sense anything produced by the intelligence of man?"[50]

The primeval soup is also discounted even by Oparin himself,

"Numerous calculations of a thermodynamic equilibrium, which would occur when the organic substances dissolved in the primordial hydrosphere were irradiated with ultraviolet or were provided with some other energy sources, show that no primeval broth could exist there, since under ultraviolet light, decomposition processes prevail over synthetic ones."[51]

D.E. Hull also comments in his *Nature* paper on the work done by Miller and others the following,

> "The conclusion from these arguments presents the most serious obstacle, if indeed it is not fatal, to the theory of spontaneous generation. First, thermodynamic's calculations predict vanishingly small concentrations of even the simplest organic compounds. Secondly, the reactions that are invoked to synthesize such compounds are seen to be much more effective in decomposing them. The physical chemist, guided by the proved principles of chemical thermodynamics and kinetics, cannot offer any encouragement to the biochemist who needs an ocean full of organic compounds to form even lifeless coacervates."[52]

Occasionally, there may be some who contend that cloud cover sheltered the early Earth from the harmful ultraviolet light. Theoretically, lightning discharges could have provided the required energy for the formation of the organic compounds. Geophysicist, Louis Lerman, has also come up with the idea that the original chemical reactions that led to the production of living organisms took place in the ocean's bubbles. The Holt 1996 text states emphatically that,

> "....inside the bubbles, the methane and ammonia required to produce amino acids would have been protected from destruction by ultraviolet light."[53]

These ideas are devoid of scientific support and are contrary to the known facts of nature with regard to the penetrating capabilities of ultraviolet light. If it were not for the presence of our ozone layer, ultraviolet light would not only penetrate the cloud cover, but would penetrate into the ocean to a depth of 60 feet. Bubbles on the surface of the ocean would be easily penetrated by ultraviolet light.

The ocean currents constantly bring water from the bottom of the ocean to the surface. Consequently, any organic compounds which had descended into the ocean would be destroyed by the deep penetrating ultraviolet light.

It has also been acknowledged that the mild electrical discharges that were used in the Miller experiment do not even come near to emulating real lightning, which includes destructive temperatures up to 50,000 degrees Fahrenheit. Research shows that,

"...if actual lightning struck rather than the fairly mild discharges used by Miller, any organics that happened to be present could not have survived"[54]

Further, biochemist, H. R. Hulett, reports,

"It is in fact hard to reconcile the thermodynamic and kinetic characteristics of these compounds with the postulated pathways for chemical evolution in the primitive environment."[55]

Wrong Results

The students are led to believe that because of Miller's experiment, amino acids, the building blocks of proteins, could have formed by chance processes in the distant past. The textbook states,

"These results demonstrated that some basic chemicals of life could have formed spontaneously on the early Earth under conditions like those in the experiment."[56]

However, the students are not informed of the fact that Miller's product was totally useless with regard to the formation of proteins and living cells. According to Cyril Ponnamperuma, a scientist in chemical evolution and the origin of life, amino acids come in two forms,

"The amino acids found in proteins of living organisms are all levorotatory [left-handed] or L-amino acids, and the sugars are dextrorotatory [right-handed] or D-sugars..."[57]

The left-handed and right-handed amino acids are identical in their atomic composition, but they differ in their three dimensional arrangement of their atoms. For instance, your two hands are identical in their composition, but they are not superimposable. That is why you can't put your right hand in your left glove. The problem for Miller's experiment is that proteins in living cells are made of only left-handed amino acids. Miller's and all other laboratory experiments, however, produced about 50% left-handed and 50% right-handed amino acids. If one of the long, right-handed forms became inserted in the short, left-handed chain, proper protein function would be effectively prevented.

With all of our expensive scientific laboratories and equipment, researchers are still mystified by the unexpected order in which amino acids in living things are found. The British chemist, Leslie Orgel, asked the questions for which, from the evolutionary point of view, there are no answers,

> "First, why do all amino acids in proteins or all nucleotides in nucleic acids have the same handedness. Secondly, why are the amino acids all left-handed (L-) and the nucleotides all right-handed (D)? We do not know the answer to either question..."[58]

Dr. Dose, the director of the Institute for Biochemistry in Johaness Gutenberg University, West Germany, emphasizes the mystery of left-handedness and right-handedness in various molecules. [Chirality refers to the fact that sugar molecules in DNA and RNA twist to the right and amino acids, the building blocks of proteins, twist to the left],

> "In spite of many attempts, there have been no breakthroughs during the past 30 years to help explain the origin of chirality in living cells."[59]

It has been postulated that the amino acids in living organisms are all left-handed because naturalistic processes operate under an extreme bias which results in the production of only the left-handed forms. However, experiments have shown repeatedly that no such bias exists. The late C. Patterson of the California Institute of Technology writes,

> "If a sugar or an amino acid is chemically synthesized in a test-tube, the product will be a mixture of left-handed and right-handed molecules, in roughly equal numbers. But in living organisms of every sort, all the amino acids are L amino acids, and in nucleic acids all the sugars are D-molecules."[60]

Science correspondent, Jon Cohen, reports on the importance of amino acid's left-handedness as underscored by the research of organic chemist William Bonner,

> "Bonner argued that homochirality is essential for life because without it, genetic material could not copy itself. Specifically, studies have shown that the two complementary strands of genetic

material that make up DNA cannot bind with each other if they are in a 'racemic' mixture, a state in which there is an equilibrium of left-handed and right-handed enantiomers."[61]

Dr. Massimo Pigliucci of the University of Tennessee relates,

"Miller put the whole thing in a ball, gave it an electric charge, and waited. He found that amino acids and other fundamental complex molecules were accumulating at the bottom of the apparatus. His discovery gave a huge boost to the scientific investigation of the origin of life. Indeed, for some time it seemed like creation of life in a test tube was within reach of experimental science. Unfortunately, such experiments have not progressed much further than the original prototype, leaving us with a sour aftertaste from the primordial soup."[62]

Further attention to Miller's experiment show the "building blocks" were not as substantial as had previously been assumed. Trevor Palmer of Nottingham Trent University, UK, states,

"The difficulties start to become clear when we look to see what building blocks were actually formed in the course of the simulation experiments of Miller and others. Robert Shapiro, professor of chemistry at New York University, pointed out in his 1988 book, *Origins*, that, contrary to what is sometimes stated, there was no synthesis of nucleotides and, indeed, only two of the component bases (adenine and guanine) were produced to any significant extent. Also, only two of the required 20 amino acids were formed in appreciable amounts, which was reduced to one when the experiment was repeated later with a less reducing atmosphere, reflecting changing views about the primeval Earth."[63]

The Holt 1994 text states on page 202 under the title, "Origin of the First Cells," that, "Most scientists think DNA evolved after simple cells had arisen." The student is thus required to think that there are simple cells and complex cells. The "simple cells", of course, theoretically evolved from proto-cellular material into complex cells. But what are simple cells? They are simply imaginary. Green and Goldberger, in their book, *Molecular Insights into the Living Process*, point out that,

"The popular conception of primitive cells as the starting point for the origin of the species is really erroneous. There was nothing functionally primitive about such cells. They contained basically the same biochemical equipment as do their modern counterparts."[64]

"How then, did the precursor cell arise? The only unequivocal rejoinder to this question is that we do not know."[65]

Further, these two researchers emphasize the greater challenge of getting from macromolecules to cells,

"Although seven steps are shown, leading from atoms to ecosystems, there is one step that far outweighs the others in enormity: the step from macromolecules to cells. All the others can be accounted for on theoretical grounds--if not correctly, at least elegantly. However, the macromolecule-to-cell transition is a jump of fantastic dimensions, which lies beyond the range of testable hypothesis. In this area all is conjecture. The available facts do not provide a basis for postulating that cells arose on this planet"[66]

Jacques Monod, in his book, *Chance & Necessity*, states,

"....we have no idea what the structure of a primitive cell might have been. The simplest living system known to us, the bacterial cell....in....its overall chemical plan is the same as that of all other living beings. It employs the same genetic code and the same mechanism of translation as do, for example, human cells. Thus the simplest cells available to us for study have nothing 'primitive' about them....no vestiges of truly primitive structures are discernible."[67]

In a word, the Miller-type experiments are totally inadequate to explain the origin of life by chance processes. New York University chemist, Robert Shapiro, emphasizes this point,

"By extrapolation of these results, some writers have presumed that all life's building blocks could be formed with ease in Miller-type experiments and were present in meteorites. This is not the case. Amino acids, such as those produced in experiments like Miller's, are far less complex than nucleotides."[68]

"The amino acids and other substances that were prominent in the Miller experiment contained two and three carbon atoms. In contrast, no nucleotides of any kind have been reported as products of spark-discharge experiments or in studies of meteorites."[69]

The long and short of all this is that the theories presented in the state-adopted high school biology texts are not supported by evidence from nature, nor from laboratory experiments. The sophistication of cellular (genetic/DNA) components is much greater than anyone anticipated, and it is infinitely more complex than the simple glob of protoplasm that Huxley, Darwin's bulldog, envisioned. In a *Scientific American* article, Robert Wagner & Miroslav Radman, researchers in the area of biotechnology, report,

"The set of genetic instructions for humans are roughly three billion letters long...even the DNA of bacterium is highly complex, containing at least three million units, all arranged in a very precise meaningful sequence."[70]

A leading researcher, Leslie Orgel, admits the failure of chemical evolution to play a significant part in the origin of life,

"It is extremely improbably that proteins and nucleic acids, both of which are structurally complex, arose spontaneously in the same place at the same time. Yet it also seems impossible to have one without the other. And so, at first glance, one might have to conclude that life could never, in fact, have originated by chemical means."[71]

Actually, the more research that is done, the more it becomes apparent that life could not have begun by chemical means or by any purely naturalistic processes. As Fred Hoyle, the english astronomer, and his colleague C. Wickramasinghe, relate,

"No matter how large the environment one considers, life cannot have had a random beginning... The trouble is that there are about two thousand enzymes, and the chance of obtaining them all in a random trial is only one part in 1,000,000... this simple calculation wipes the idea entirely out of court...The enormous information content of even the simplest living systems...cannot in our view be generated by what are often called "natural" processes..."[72]

In fact, molecular biologist, Francis Crick, and his colleague, Leslie Orgel, acknowledge that the origin of living cells is in the realm of the miraculous,

> "...the origin of life appears at the moment to be almost a miracle, so many are the conditions which would have had to have been satisfied to get it going."[73]

John Horgan, the former senior writer at *Scientific American*, says,

> "Don't tell the creationists, but scientists don't have a clue how life began,"[74]

He goes on to say,

> "Dennis Overbye just wrote a status report for *The New York Times* on research into life's origin, based on a conference on the topic at Arizona State University. Geologists, chemists, astronomers and biologists are as stumped as ever by the riddle of life."[75]

> "More than 30 years of experimentation on the origin of life in the fields of chemical and molecular evolution have led to a better perception of the immensity of the problem of the origin of life on earth rather than to a solution. At present, all discussion on principle theories and experiments in the field either end in a stalemate or a confession of ignorance." (Dr. Klaus Dos, Head of the Institute for Biochemistry at the Gutenberg University of West Germany)

Surely this vast cellular civilization involving the amazing complexity of the whole DNA/genetic system (since it bears the effects that would ordinarily be attributed to an extremely intelligent designer) should, if science is going to retain the element of integrity, be attributed to an intelligent designer.

Chapter One End Notes

---

[1] George B. Johnson, Biology: Visualizing Life, Holt, Rinehart & Winston, 1994, p. 201
[2] Miller & Levine, Biology, Prentice Hall, 1995, p. 313
[3] Biology The Dynamics of Life, Macmillan/McGraw Hill, p. 412
[4] George B. Johnson and Peter H. Raven, Biology: Principles and Explorations, Holt, Rinehart and Winston, Inc. 1996, p. 227
[5] Albert Towle, Modern Biology, Holt, Rinehart & Winston, 1989, p. 211
[6] Alton Biggs, Dorcas C. Green, Kathleen Gregg, Whitney Crispen Hagins, Chris Kapicka, Linda Lundgren, Peter Rillero, & the National Geographic Society, *Biology: The Dynamics of Life*, Glencoe/ McGraw-Hill, 2000, p. 391
[7] George B. Johnson and Peter H. Raven, Biology: Principles and Explorations, Holt, Rinehart and Winston, Inc. 1996, p. 227
[8] Bernice Essenfeld, Carol Gontang, Randy Moore, Biology, Addison-Wesley, 2000, p.258
[9] Ibid.
[10] George B. Johnson, Biology Visualizing Life, Holt, Rinehart & Winston, 1994, p. 112
[11] Kenneth R. Miller & Joseph Levine, Biology, Pearson/Prentice Hall, 2004, p. 424
[12] Kenneth R. Miller and Joseph Levine, Biology, Prentice Hall, 1998, p. 344
[13] Ibid.
[14] Bernice Essenfeld, Carol Gontang, Randy Moore, Biology, Addison-Wesley, 1994, p. 226
[15] Ibid.
[16] George B. Johnson, Biology Visualizing Life, Holt, Rinehart & Winston, 1994, p. 201
[17] Time, October 11, 1993
[18] Time, October 11, 1993
[19] P. Abelson, "Chemical Events on the Primitive Earth," Proceedings of the National Academy of Science, 1966 June 55(6), p. 1365
[20] Ibid.
[21] Steven Johnson, *The Invention of Air: A Story of Science, Faith, Revolution, and the Birth of America*, Riverhead Hardcover, 2008
[22] C.F. Davidson, "Geochemical Aspects of Atmospheric Evolution," Proceedings of the National Academy of Science, 53(1956): 1203
[23] Michael Denton, Evolution: A Theory in Crisis, London: Burnett Books, Ltd., 1985: p. 261
[24] Erich Dimroth and Michael M. Kimberly, 1976, "Precambrian Atmospheric Oxygen: Evidence in the Sedimentary Distribution of Carbon, Sulfur, Uranium and Iron," in *Canadian Journal of Earth Sciences*, 13(9): p. 1161-1185
[25] H. Clemney and N. Badham, Geology, vol. 10 (1982), p. 141
[26] Patrick Huyghe, "New Recipe For Cosmic Soup," Science Digest, May 1983, p. 44
[27] _____, "New Evidence on Evolution of Early Atmosphere and Life", *Bulletin of the American Meteorological Society*, vol. 63, November 1982, pp. 1329
[28] H. Clemmey & Nick Badham "Oxygen in the Precambrian Atmosphere: An Evaluation of the Geological Evidence" Geology, Vol. 10, March 1982, p. 141
[29] *Ibid.*
[30] Monitor, "Smaller Planets Began with Oxidized Atmosphere," New Scientist Vol. 87, July 10, 1980, p. 112
[31] Marcel Florkin, "Ideas and Experiments in the Field of Prebiological Chemical Evolution," Comprehensive Biochemistry 29B, 1975, p. 231-260
[32] Molecular Evolution and the Origin of Life, Revised Edition, Marcel Dekker, 1977, p. 43, 74-76
[33] Michael Denton, Evolution: A Theory in Crisis, Burnett Books, Ltd., 1985, pp. 261-262
[34] Carl Sagan, *Journal of Theoretical Biology*, vol. 39, p. 197
[35] Francis Hitching, The Neck of the Giraffe - Darwin, Evolution and the New Biology, New American Library, New York, 1982, p. 48
[36] *Ibid.*
[37] Biology: The Dynamics of Life,[Macmillan/McGraw Hill, p. 412

[38] Harold Blum, *Time's Arrow and Evolution*, 1968, p. 158
[39] Richard Milner, Encyclopedia of Evolution, p. 274
[40] Michael Denton, Evolution: A Theory in Crisis, London: Burnett Books, Ltd., 1985, p. 261
[41] Noam Lahav, *Biogenesis: Theories of Life's Origins*, Oxford University Press, 1999, pp. 138-139
[42] Bernice Essenfeld, Carol Gontang, Randy Moore, Biology, Addison-Wesley, 1994, p. 227
[43] Freeman Dyson, "Honoring Dirac," Science, vol. 185 [September 27, 1974], pp. 1160-1161
[44] Fred Hoyle, The Intelligent Universe, 1983, p. 23
[45] Jeffrey Bada & Antonio Lazcano, "Prebiotic Soup", Science, May 2, 2003, p. 746
[46] Fisher, Grolier Multimedia Encyclopedia, 2003, fossil section
[47] _____, "New Evidence on Evolution of Early Atmosphere and Life", Bulletin of the American Meteorological Society, vol. 63, November 1982, p. 1329
[48] S.L. Miller and H.C. Urey: "Organic Compound Synthesis on the Primitive Earth", Science, 130 (1959) p. 247
[49] D.E. Hull: "Thermodynamics and Kinetics of Spontaneous Generation", Nature, 186 (1960) pp. 693-695
[50] Michael Denton, Evolution, A Theory in Crisis, 1985, pp.327-8, 342.
[51] A. Oparin, "Problem of the Origin of Life: Present State and Prospects," *Chemical Evolution and the Origin of Life*, Eds. R. Buver & C. Ponnamperuma, 1971, p. 6
[52] D. E. Hull, "Thermodynamics and Kinetics of Spontaneous Generation", in Nature, 186 (1960) pp. 693-695
[53] George B. Johnson & Peter H. Raven, Biology: Principles and Explorations, Holt, Rinehart & Winston, Inc. 1996, p. 229
[54] *Science News*, Dec. 1, 1973: p. 340
[55] H.R. Hullett, Journal of Theoretical Biology, Vol. 24, p. 70, 1969
[56] George B. Johnson & Peter H. Raven, *Biology: Principles and Explorations,* Holt, Rinehart & Winston, Inc. 1996, p. 228
[57] C. Ponnamperuma, The Origins of Life, 1972, pp. 105-06
[58] Leslie Orgel, "Darwinism at the Very Beginning of Life," New Scientist 94, Apr. 15, 1982: p 151
[59] Dose, Professor Dr. Klaus, "The Origin of Life; More Questions than Answers," Interdisciplinary Science Reviews, vol. 13, no. 4, [1988], p. 352
[60] C. Patterson, *Evolution*, 1978, p. 157
[61] Jon Cohen, "Getting All Turned Around Over the Origins of Life on Earth," *Science*, vol. 267, [March 3, 1995], p. 1265
[62] Massimo Pigliucci, "Where Do We Come From?" *Skeptical Inquirer*, Vol. 23, September/October 1999, p.24
[63] Trevor Palmer, *Controversy Catastrophism & Evolution: The Ongoing Debate*, Kluwer Academic/ Plenum Publishers, 1999, p. 266
[64] David E. Green and Robert F. Goldberger, *Molecular Insights into the Living Process*, Academic Press, 1967, p. 403
[65] *Ibid.*
[66] *Ibid.*, pp. 406-407
[67] Jacques Monod, *Chance and Necessity: An Essay on the Natural Philosophy of Modern Biology*, Knopf, 1971, p. 134
[68] Robert Shapiro, "A Simpler Origin for Life," *Scientific American*, June 2007, p. 43
[69] Ibid., p. 49
[70] Miroslav Radman and Robert Wagner, *Scientific American*, The High Fidelity of DNA Duplication, Vol. 259, No. 2, August 1988, pp. 40-46
[71] Leslie Orgel, "The Origin of Life on the Earth", Scientific American, vol. 271, October 1994, p. 78
[72] Fred Hoyle and Chandra Wickramasinghe, *Evolution from Space*, London: J.M. Dent and Sons, 1981
[73] Francis Crick and L.E. Orgel, Icarus Directed Panspermia, 1973, 19:341-46
[74] John Horgan, "Pssst! Don't tell the Creationists, but Scientists don't Have a Clue How Life Began," *Scientific American* (blog), February 28, 2011
[75] Ibid.

HOMOLOGOUS STRUCTURES

Introduction

There are two different ways of looking at structures in various animals that have similarities to each other. First, one may regard them as having been formed by a common designer or according to a common plan (Creation), or one may regard them as having been descended from a common ancestor (evolution).

Darwin wrote in the *Origin of Species*,

> "Finally, the several classes of facts which have been considered in this chapter, seem to me to proclaim so plainly, that the innumerable species, genera and families, with which this world is peopled, are all descended, each within its own class or group from common parents, and have all been modified in the course of descent, that I should without hesitation adopt this view, even if it were unsupported by other facts or arguments."[1]

Today's modern textbooks continue to promote Darwin's thinking on this subject as one of the prime examples of the "mechanisms" of evolution.

Textbook Evolution

Here and in the other categories which we address, the reader will find that several textbooks may be listed. When they are, this is for the purpose of illustrating the point that the textbooks are saying the same thing year after year, and all of the different textbook companies are essentially saying the same thing.

> "In such species we very often see **homologous structures**. These are the same body parts that have been modified in different ways in different lines of descent from a common ancestor (*homo-* means "same"). For example, most land vertebrate have homologous structures and probably share a common ancestor that had four five-toed limbs. The limbs diverged in form and became wings in pterosaurs, birds, and bats (Figure 23.7). All these wings are

homologous--they have the same parts. The five-toed limb also evolved into the flippers of porpoises and the anatomy of your own forearms and fingers."[2] (Cecie Starr & Beverly McMillan, *Human Biology*, ninth edition, Brooks/Cole, Cengage Learning) (College Textbook)

"Homologous structures are similar structures in different species that have the same general function indicating that they have been derived from a common ancestor."[3] (*Concepts in Biology,* 13th edition, McGraw-Hill)

"Evolutionary biologists view homologous structures as evidence that organisms evolved from a common ancestor."[4] (Alton Biggs, Kathleen, Gregg, Whitney Crispen-Hagins, Chris Kapicka, Linda Lundgren, Peter Rillero, *Biology: The Dynamics of Life,* Glencoe, McGraw-Hill)

"The presence of homologous features in different species indicates that the species shared a fairly recent common ancestor."[5] (*Modern Biology*, Holt, Rinehart and Winston)

"Homologous structures are one type of evidence for the evolution of living things."[6] (K.R. Miller, J. Levine, *Biology,* Prentice Hall)

"Biologists define homologous structures as anatomical structures that occur in different species and that originated by heredity from a structure in the most recent common ancestor of the species."[7] (John H. Postlethwait & Janet L. Hopson, *Modern Biology*, Holt, Rinehart, & Winston)

"Figure 8 shows that the bones of the front limbs of dolphins, birds, and dogs are arranged in a similar way. These similarities provide evidence that these three organisms all evolved from a common ancestor. Similar structures that related species have inherited from a common ancestor are known as homologous structures."[8] (Michael Padilla, et al., *California Focus on Life Science*, Pearson Prentice Hall)

"Anatomically similar structures inherited from a common ancestor are called homologous structures."[9] (Alton Biggs, et al., *Biology,* Glencoe)

Science Versus

In the 1800s when the concept of special creation by the word of God was being replaced by evolution through totally naturalistic processes, the perception of similarities between certain creatures such as human arms, bird wings and dolphin fins, became an important point of argument. For those who were confident of the Genesis account of creation, the similarities were understood as being the result of having a common designer. In the great rebellion against God, the concept of a common designer was replaced by the concept of common ancestor. Then, shortly thereafter, the science of genetics began to be developed by Gregor Mendel. Scientists involved in this argument immediately recognized that, due to the influence of genes on the various organs in living things, it would be necessary to be able to demonstrate that there was a similarity of genes, which corresponded to the similar appendages (arm, wing, fin). However, as the study of genetics progressed, it became abundantly clear to the researchers that their presumptions of identical genes were to be seriously disappointing. This means, of course, that the idea of homologous structures, as being the result of a common ancestor, would be seriously undermined by the science of genetics. The British evolutionary embryologist, Sir Gavin DeBeer, has been extremely emphatic on this subject,

> "It is now clear that the pride with which it was assumed that the inheritance of homologous structures from a common ancestor explained homology was misplaced; for such inheritance cannot be ascribed to identity of genes. The attempt to find homologous genes has been given up as hopeless."[10]

> "What mechanism can it be that results in the production of homologous organs, the same "patterns," in spite of their not being controlled by the same genes? I asked that question in 1938, and it has not been answered."[11]

In his book, *The Icons of Evolution*, Jonathan Wells states,

> "The lack of correspondence between genes and structures is true not only for entire organisms, but also for limbs."[12]

It is now commonly understood that homologous organs are not produced by homologous genes as British parapsychologist and biology teacher, John Randall, relates,

> "Homologous organs are now known to be produced by totally different gene complexes in the different species. The concept of homology in terms of similar genes handed on from a common ancestor has broken down..."[13]

The importance of this concept to the theory of evolution is explained by Michael Denton,

> "With the demise of any sort of straight forward explanation for homology one of the major pillars of evolution theory has become so weakened that its value as evidence for evolution is greatly diminished. The breakdown of the evolutionary interpretation for homology cannot be dismissed as a triviality and casually put aside as a curiosity for, as Sir Alister Hardy reminds us in his book, The Living Stream: 'The concept of homology is absolutely fundamental to what we are talking about when we speak of evolution-yet in truth we cannot explain it all in terms of present day biological theory.'"[14]

Jonathan Wells (as quoted by Lee Strobel) further expounds on the problem for the theory of evolution,

> "A more common explanation nowadays is that the homologies come from similar genes. In other words, the reason two features are homologous in two different animals would be that they're programmed by similar genes in the embryo. But it turns out this doesn't work very well, either. We know some cases where you have similar features that come from different genes, but we have lots and lots of cases where we have similar genes that give rise to very different features."[15]

Jonathan Wells also reveals that this problem for evolution has been known for a very long time,

> "...biologists have known for a hundred years that homologous [similar] structures are often not produced by similar

developmental pathways. And they have known for thirty years that they are often not produced by similar genes, either. So there is no empirically demonstrated mechanism to establish that homologies are due to common ancestry rather than common design."[16]

With regard to the battle over the status of *Ramapithecus* as a hominid, there occurred a debate over the relative importance of discerning relationships between molecular evidence and fossil evidence. After Roger Lewin in his book, *Bones of Contention*, refers to comments by Adrienne Zihlman, he goes on to say,

> "The key issue is the ability correctly to infer a genetic relationship between two species on the basis of a similarity in appearance, at gross and detailed levels of anatomy. Sometimes this approach works, but sometimes it can be deceptive, partly because similarity of structure does not necessarily imply an identical genetic heritage: a shark (which is a fish) and a porpoise (which is a mammal) look similar because they have become adapted to the same environment, not because they are close genetic cousins."[17]

The homologous structures are listed as evidence for evolution in the high school and college text. However, in order to make this case, investigation must be made on the genetic level to verify that there are also homologous genes. Once this idea is subjected to careful scientific scrutiny, the case for evolution in this area falls apart. This is clearly related by Stephen Meyer of the Discovery Institute,

> "In sharks, for example, the gut develops from cells in the roof of the embryonic cavity. In lampreys, the gut develops from cells on the floor of the cavity. And in frogs, the gut develops from cells from both the roof and the floor of the embryonic cavity. This discovery--that homologous structures can be produced by different developmental pathways--contradicts what we would expect to find if all vertebrates share a common ancestor....To summarize, biologists have made two discoveries that challenge the argument from anatomical homology. The first is that the development of homologous structures can be governed by different genes and can follow different developmental pathways. The second discovery, conversely, is that sometimes the same gene plays a role in

> producing different adult structures. Both of these discoveries seem to contradict neo-Darwinian expectation."[18]

Lewin's remarks are correct and certainly should be applied to the concept of homology with regard to the wing of a bird, the fin of a dolphin, and the arm of a human,

> "I think students deserve to know that scientists haven't resolved this problem. Instead, some textbooks simply define homology as similarity due to common ancestry. So the theory becomes true by definition. What the textbook is saying is that similarity due to common ancestry is due to common ancestry. And *that's* circular reasoning."[19]

Chapter Two End Notes

[1] Charles Darwin, *The Origin of Species*, Vol. XI, The Harvard Classics, P.F. Collier & Son, 1909-14

[2] Cecie Starr & Beverly McMillan, *Human Biology*, ninth edition, Brooks/Cole, Cengage Learning, 2010, p. 450) (College Textbook)

[3] Eldon D. Enger, Frederick C. Ross, & David B. Bailey, *Concepts in Biology,* 13th edition, McGraw-Hill, 2009, p. 299

[4] Alton Biggs, Kathleen, Gregg, Whitney Crispen-Hagins, Chris Kapicka, Linda Lundgren, Peter Rillero, *Biology: The Dynamics of Life,* Glencoe, McGraw-Hill, 2000, pp.408-409

[5] *Modern Biology*, Holt, Rinehart and Winston, 2002, p. 290

[6] K.R. Miller, J. Levine, *Biology,* Prentice Hall, 2006, p. 384

[7] John H. Postlethwait & Janet L. Hopson, *Modern Biology*, Holt, Rinehart, & Winston, 2006, p. 305

[8] Michael Padilla, et al., *California Focus on Life Science*, Pearson Prentice Hall, 2008, p. 235

[9] Alton Biggs, Whitney Crispen Hagins, William G. Holliday, Chris L. Kapicka, Linda Lundgren, Ann Haley MacKenzie, William D. Rogers, Marion B. Sewer, Dinah Zike, *Biology,* Glencoe, 2007, p. 424

[10] Sir Gavin Debeer, *Homology an Unsolved Problem*, Oxford University Press, 1971, p. 16

[11] *Ibid.*

[12] Jonathan Wells, *Icons of Evolution*, Regnery Publishing, Inc., 2000, p. 74

[13] John Randall, *Parapsychology and the Nature of Life*, Harper & Row, 1977 p. 210

[14] Michael Denton, *Evolution: A Theory in Crisis*, Maryland: Adler and Adler Publishers, 1986, p. 151

[15] Lee Strobel quoting Jonathan Wells in *The Case for a Creator*, Zondervan, 2004, p. 53

[16] Jonathan Wells, "Survival of the Fakest," *The American Spectator*, December 2000/January 2001, p. 22

[17] Roger Lewin, *Bones of Contention*, A Touchstone Book, Simon & Schuster, 1987, p. 123

[18] Stephen Meyer, Scott Minnich, Jonathan Moneymaker, Paul Nelson, and Ralph Seelke, *Explore Evolution: The Arguments For and Against Neo-Darwinism*, Hill House, 2007, pp. 44-45

[19] Jonathan Wells *Op. Cit.*, p. 54

MUTATIONS

Introduction

The public at large is probably not aware of a remarkable fact which the eminent Harvard biologist, Ernst Mayr, reveals,

> "Darwin never really did discuss the origin of species in his *On the Origin of Species*."[1]

He did not explain the origin of the kinds of new traits needed to produce a new kind of organism. He explained only variations of existing kinds. Variation had long been a significant part of the Creation Model. In time, Mendelian genetics was later incorporated into Darwin's theory of natural selection. The fusion of these two separate disciplines from the early 1920s through the late 1950s became known as Neo Darwinism. Neo Darwinists accept Darwin's ideas with regard to natural selection, but they are confident that new traits come about by chance, by random changes known as gene mutations. Evolutionists were forced to give up Darwin's original idea of use and disuse since it had no significant evidence to support it. It has been hoped that mutations would lend adequate scientific support to the theory of evolution. However, as will be shown in this section, mutations lend no significant support whatsoever.

Since those promoting the theory of evolution totally disregard the possibility of special creation by God, they must come up with mechanisms by which organisms can experience changes that natural selection can use to their advantage. Geneticists have come up with two general explanations for the variation that is visible in living organisms: recombination and mutation.

Recombination is merely the reshuffling of genes that are already present in the organism. It can provide the offspring with a new combination of traits, some of which were concealed by dominant genes in other combinations and are now expressed in the new combinations.

Mutation provides the organism with changes that are inheritable in the genes themselves. Mutations, which can be caused by chemical radiation, represent mistakes in the biological process.

Recombination can be no help to the theory of evolution. The recombination of genes present in a given population can indeed be used to explain the variations which we see. For instance, there are about 37 varieties of flying squirrels. They are, however, all easily identified as flying squirrels. None are in the process of becoming a bat or something else.

In fact, it is this business of recombination that makes it possible for certain kinds of bacteria to become resistant to various antibiotics. It is the reshuffling of genes, not evolution, that brings about their resistance.

To use recombination to explain variation means that natural selection can only occur from traits that were already present in the genes. This phenomenon has always been a significant part of the Creation model. As a part of the Evolution model, recombination can not be shown to lead to a higher organism. Evolutionists believe that the earlier organisms were simple. In order to get the kinds of changes that natural selection can choose to preserve that might actually lead to a higher species, evolutionists have had to rely most heavily on mutations. However, repeated research and experimentation with mutations has not been able to show that mutations could have indeed provided the mechanism needed for advancing a simpler species to a more complex one.

The strategy employed by biology textbook writers is to convince the students that mutations are real and then induce them to believe that species have advanced over time by means of positive mutations. There are, of course, no known positive mutations when positive mutations are properly explained. Certain mutations have been called positive, but under examination they are found to be only "equivocally" positive. In the realm of "positive" mutations are two kinds: "equivocally positive" and "unequivocally positive." "Equivocally positive" mutations always have a downside that pose danger to the organism. "Unequivocally positive" mutations have no downside and are entirely beneficial. There are no "unequivocally positive" mutations. The textbooks, when referring to "positive" mutations are always talking about mutations that are "equivocally" positive. If the textbook actually cites a positive mutation (equivocal), they always cite the same one because there is only one that is known to occur in human beings--sickle cell anemia. The students are not well informed about the problems caused by mutations. If they were, at least some of them would wonder how this mechanism could actually benefit any organism. They are not told that there are now almost 4000 diseases caused by mutations.

"The human genome contains a complete set of instructions for the production of a human being.... Genome research has already exposed errors |mutations| in these instructions that lead to heart disease, cancer, and neurological degeneration."[2]

Textbook Evolution

"A change in the gene will produce a change in the offspring of an organism. Changes in genes are called mutations. And mutations are one of the driving forces behind evolution."[3] (*Evolution: Change Over Time*, Prentice Hall)

"Scientists now know that new variations come from two sources: mutation and recombination."[4] (Bernice Essenfeld, Carol Gontang, Randy Moore, *Biology*, Addison-Wesley)

"Mutations in homeobox genes could explain the sudden appearance of new traits in an individual and eventually in entire populations."[5] (*Ibid.*)

"Flies with homeobox mutations have such strange traits as extra pairs of wings and legs growing out of the head!"[6] (*Ibid.*)

"Sickle Cell Anemia reveals Evidence of Microevolution"[7] (Alton Biggs, Chris Kapicka, Linda Lundgren, *Biology, The Dynamics of Life*, Glencoe, McMillian/McGraw-Hill)

"Mutations are important in evolution because they add new genetic material to the gene pool. Many random mutations are harmful and most have no significance at all. However, once in a great while, they provide a favorable variation to an offspring, and natural selection takes over to distribute this variation to the following generations."[8] (*Ibid.*, p. 437)

"Mutations are the raw material for evolutionary change; without mutations there could be no new variations among members of a population."[9] (Sylvia S. Mader, *Biology,* McGraw Hill, 2001, p.304)

In order for one species to change into another the genes must undergo change. The only way these changes can happen is by the well-known concept of mutations, as Mayr explains,

> "It must not be forgotten that mutation is the ultimate source of all genetic variation found in natural populations and the only new material available for natural selection to work on."[10]

After many years of observing mutations that occur naturally and mutations which have been forced on organisms by artificial means, it has become obvious that they continually have a negative effect on the organism and have not been found to be capable of producing improvements, let alone a new organism all together.

First, let us deal briefly with the textbook statements promoting the idea that sickle-cell anemia is good evidence for microevolution. The reason evolutionists make this argument is because people in Africa who have sickle cell anemia are resistant to malaria. Thus, they say this is a positive mutation which is helpful to the individual. This is indeed a devious argument because sickle cell anemia is a disease, which by itself, can cause much suffering and even death.

The health situation in Africa is burdened by the fact that the disease, malaria, causes the death of about two million people each year. The cause of this disease has to do with mosquitos. Female mosquitoes of the species anopheles are able to transmit the malaria parasite from one human to another.

There is also another problematic health condition prevalent in Africa known as sickle cell anemia. This is due to a mutation in a hemoglobin gene. If a baby inherits two hemoglobin genes (one from each parent) with the sickle cell mutation, it will become ill and very often dies within several years. If the baby inherits only one gene with the sickle cell mutation it will not die, but it will be beset with difficult and challenging health conditions. As a result of their "sickle" or distorted shape, these cells deliver less oxygen to the body than do the healthy disc shaped cells. This results in a wide variety of health problems. Among these, the most common is anemia and the occurrence of pain attacks due to the plugging of the blood vessels. The pain attacks may last for shorter or for longer periods of time. Further, the attacks may be mild in discomfort or they be very severe. Besides the most common problems (anemia and pain attacks), other challenging problems may also develop such as strokes, pneumonia, gall stones and growth retardation during childhood.

At present there is no cure for the disease of sickle cell anemia. The difficulties that are associated with this health problem must simply be managed by a variety of treatments which include pain relievers, transfusions, antibiotics and surgery.

A person who has this condition suffers from a challenging blood cell problem. Ordinarily, the red blood cells in a human are round or disc shaped. By contrast, many of the blood cells in people who have sickle cell anemia are deflated in shape, or "sickled", and are unable to carry the normal amount of oxygen to the various tissues of the body, which results in anemia. As a result of their sickle shape, these cells are fragile and break down so quickly that they can not be adequately replaced. Consequently, anemia and distortion of normal blood flow are responsible for a variety of health problems endured by these individuals.

The protozoan single celled parasite is spread by the female mosquito, attempting to obtain from the blood important protein that is necessary for her developing eggs. After getting inside the person, the malaria parasites get into the person's red cells where they are able to evade the circulating serum antibodies. There, they also eat the hemoglobin, the red protein that carries oxygen in the cell. During this period they also undergo asexual reproduction. At the right time of the malaria life cycle, the red blood cell is caused to burst. This, in turn, releases hundreds of new parasites, plus the toxic products of metabolism. It is these waste products which are responsible for the chills and fevers that are associated with malaria.

It is obvious that no researcher on his way to study diseases in Africa would volunteer to have himself bit by these mosquitos in order to ensure that he would not come down with malaria. It is equally obvious that this kind of mutation could have no part in improving any species, let alone be useful in the development of an entirely new species.

The quotes listed below are from a variety of competent scientists, all evolutionists, whose research supports the above statement.

1. Mutations are rare.

Francisco Ayala, who is with the Department of Ecology and Evolutionary Biology at the University of California, Irvine, states,

> "Although mutation is the ultimate source of all genetic variation, it is a relatively rare event...."[11]

He also says,

> "It is probably fair to estimate the frequency of a majority of mutations, in higher organisms, between one in ten thousand and one in a million per gene per generation."[12]

The English Scientist, Richard Dawkins, also relates that,

> "In real life, the probability that a gene will mutate [that is, in one generation] is often less than one in a million."[13]

D. Erwin and J. Valentine of the University of California indicate the improbability of consecutive mutations occurring in the same area,

> "Variable mutations with the major morphological or physiological effects are exceedingly rare and usually infertile; the chance of two identical rare mutations individuals arising in sufficient propinquity to produce offspring seems too small to consider as a significant evolutionary event."[14]

Geneticist John Endler of the University of California indicates that the relationship of mutations to evolution is unclear,

> "Although much is known about mutation, it is still largely a 'black box' relative to evolution. Novel biochemical functions seem to be rare in evolution, and the basis for their origin is virtually unknown."[15]

The facts reported by *World Book Encyclopedia* should clarify to the objective reader that any confidence regarding the value of mutations to the theory of evolution is unwarranted,

> "Mutations rarely occur. Most genes mutate only once in 100,000 generations or more. Researchers estimate that a human gene may remain stable for 2,500,000 years."[16]

Francisco Ayala also states,

> "It therefore seems clear that, contrary to Darwin's conception, most of the genetic variation in populations arises not from new mutations at each generation but from the reshuffling of previously accumulated mutations by recombination."[17]

Before we go further with this comment, let us stop to recognize that recombinations allow for variation *within* a species and does not lead to macroevolution--from one species to a higher species. He continues,

> "Although mutation is the ultimate source of all genetic variation, it is a relatively rare event, providing a mere trickle of new alleles into the much larger reservoir of stored genetic variation. Indeed recombination alone is sufficient to enable a population to expose its hidden variation for many generations without the need for new genetic input by mutation."[18]

2. Mutations are mistakes.

In an article in *Science* magazine, the Environmental Mutagenic Society informs us that,

> "Being an error process, mutation consists of all possible changes in the genetic material (excluding recombination and segregation)."[19]

To assume that the complex genetic machinery found in all living organisms could by means of genetic errors produce over time more and more competent and complex species is not a matter of scientific observation, but of wishful thinking.

3. Mutations are not Cumulative

Pierre-Paul Grasse, who held the chair in Evolution at the University of Sorbonne in France for twenty years explains the problems that occur when scientists try to assign mutations as a useful mechanism for evolution,

> "Mutations, in time, occur incoherently. They are not complementary to one another, nor are they cumulative in successive generations toward a given direction. They modify what preexists, but they do so in disorder..."[20]

Grasse goes on to emphasize the problem of the disorder that occurs as a result of a mutation,

> "As soon as some disorder, even slight, appears in an organized being, sickness, then death follow. There is no possible compromise between the phenomenon of life and anarchy."[21]

Mutations are not only rare, but, as stated above, they are not cumulative. This means that the observable facts are that the mutations in an organism occur randomly. However, in a proposed transition, such as reptile to bird, the mutations, changing the reptilian forelegs into wings, would necessitate that the mutations over long periods of time would be cumulative. They would have to continually be taking place in the forelegs. This does not happen at all. As Grasse says, they occur randomly, that is, helter-skelter: head, leg, shoulder, stomach, etc. Grasse admits that to go against the observed way of nature in these things would demand a miraculous turn of events.

> "The opportune appearance of mutations permitting animals and plants to meet their needs seems hard to believe. Yet the Darwinian theory is even more demanding: a single plant, a single animal would require thousands and thousands of lucky, appropriate events. Thus, miracles would become the rule: events with an infinitesimal probability could not fail to occur....There is no law against day-dreaming, but science must not indulge it."[22]

As Francisco Ayala indicates, the random nature of mutations is not conducive to benefits in the present or the future,

"The forces that give rise to gene mutations operate at random in the sense that genetic mutations occur without reference to their future adaptiveness in the environment."[23]

4. Mutations are Harmful

From H.J. Muller's famous research experiments we learn,

"But mutations are found to be of a random nature, so far as their utility is concerned. Accordingly well over 99%, are harmful in some way, as is to be expected of the effects of accidental occurrences."[24]

Julian Huxley was an avid supporter of the theory of evolution. He was a major player in the modern development of Humanism, which sought to create a new religion without revelation, a non-theistic religion. His statement, "the great majority," is essentially a statement of wishful thinking on the part of a man who knew of the challenges that mutations posed to the theory of evolution, but who was holding out for the possibility that somewhere or sometime the positive mutations would be discovered. As the reader will find out in the course of this section, no such positive mutations are known,

"One would expect that any interference, with such a complicated piece of chemical machinery as the genetic constitution would result in damage. And, in fact, this is so: The great majority of mutant genes are harmful in their effects on the organism."[25]

The Russian American Scientist and prolific writer of articles promoting evolution, Theodosius Dobzhansky, also writes of the harmful effects of mutations,

"Most mutants which arise in any organism are more or less disadvantageous to their possessors. The classical mutants obtained in Drosophilia usually show deterioration, breakdown, and disappearance of some organs.... the deleterious character of most mutations seems to be a very serious difficulty."[26]

Professor James Crow, at the University of Wisconsin states,

"Mutations and mutation rates have been studied in a wide variety of experimental plants and animals, and in man. There is one general result that clearly emerges: almost all mutations are harmful. The degree of harm ranges from mutant genes that kill their carrier to those that cause only minor impairment."[27]

He also acknowledges that,

"...mutants would usually be detrimental. For a mutation is a random change of a highly organized, reasonably, smoothly functioning living body. A random change in the highly integrated system of chemical processes which constitute life is almost certain to impair it--just as random interchange of connections in a television set is not likely to improve the picture."[28]

Now, when we take a close look at one single item such as a very small change in hemoglobin, we find, as researched by the late George Wald, Professor of Biology at Harvard University, that the effect of such a mutation is indeed negative,

"...I took a little trouble to find whether a single amino acid change in a hemoglobin mutation is known that doesn't affect seriously the function of that hemoglobin. One is hard put to find such an instance."[29]

In the past, the Environmental Mutagenic Society's study of this subject acknowledges not only that mutations were harmful but they should be avoided,

"Most mutations producing effects large enough to be observed are deleterious,....Furthermore, the wide variety of mechanisms by which radiations and chemicals induce mutations make it very unlikely that generalized schemes can be devised to protect against mutagens, except by avoiding them in the first place."[30]

If you increase the mutation rate then you increase the possibilities of the organism being harmed. With regard to humans, the Society relates that,

"Since the vast majority of detectable mutations are deleterious, an artificially increased human mutation rate would be expected to be harmful in proportion to the increase."[31]

Though it is well known that mutations are harmful, it has been postulated that now and then a good, or helpful, mutation would arise and that over millions of years such helpful mutations would be able to facilitate the transition of a lower species into a higher one. C.P. Martin of McGill University indicates that such helpful mutations have not been discovered,

"The truth is that there is no clear evidence of the existence of such helpful mutations. In natural populations endless millions of small and great genetic differences exist, but there is no evidence that they arose by mutation."[32]

"For any acceptable theory of the mechanism of evolution, a great number of fully viable hereditary variations is necessary. Mutation does produce hereditary changes, but the mass evidence shows that all, or almost all, known mutations are unmistakably pathological and the few remaining ones are highly suspect."[33]

Note the strong statements made below by the authors of the text *Life Its Forms and Changes,*

"However, beneficial mutations do occur, as you shall soon see."[34]

One would think that after making this dogmatic statement that the authors would provide some good examples of positive mutations. However, the examples that they give are not examples of positive mutations,

"Molds...synthesize, the amino acids and proteins they need."[35]

"Using X rays, Beadle and Tatum produced mutants that could not synthesize certain compounds."[36]

This, of course, is a negative result. What these scientists should have been able to report as a result of their X ray experience, was that these molds were *better* able to synthesize certain compounds. This would show a positive mutation. Instead, in this statement and in the ones listed below the statements reflect the fact that the radiated molds were less efficient than they were prior to the process of radiation.

> "One irradiated strain of the mold, for example, had lost its ability to synthesize thiamine, a vitamin essential to the growth of Neurospora."[37]

> "In other words, for these mutant molds to be able to grow, thiamine had to be supplied in the medium in which they were grown."[38]

> "Otherwise they died from a lack of this vitamin."[39]

In this experience of forcing mutations, as in all others that have been tried, the organisms are not improved. Francisco Ayala explains,

> "The process of mutation ultimately furnishes the materials for adaptation to changing environments. Genetic variations which increase the reproductive fitness of a population to its environment are preserved and multiplied by natural selection. Deleterious mutations are eliminated more or less rapidly depending on the magnitude of their harmful effects. High-energy radiations, such as X-rays, increase the rate of mutations. Mutations induced by radiation are random in the sense that they arise independently of their effects on the fitness of the individuals which carry them. Randomly induced mutations are usually deleterious. In a precisely organized and complex system like the genome of an organism, a random change will most frequently decrease, rather than increase the orderliness or useful information of the system."[40]

The serious problem posed by random changes in a highly complex system is further highlighted by James F. Crow, Professor of Genetics at the University of Wisconsin,

> "Even if we didn't have a great deal of data on this point, we could still be quite sure on theoretical grounds that mutants would usually be detrimental. For a mutation is a random change of a highly organized, reasonably smoothly functioning living body. A random change in the highly integrated system of chemical processes which constitute life is almost certain to impair it--just as

a random interchange of connections in a television set is not likely to improve the picture."[41]

Above, Professor Crow highlights a serious problem regarding this subject. Below, we have Warren Weaver et al. posing a question, that from the evolutionary standpoint, is unanswerable,

"Many will be puzzled about the statement that practically all known mutant genes are harmful. For mutations are a necessary part of the process of evolution. How can a good effect-evolution to higher forms of life-result from mutations practically all of which are harmful?"[42]

The late Theodosius Dobzhansky, a highly respected Russian geneticist who did much research on fruit flies, also notes the deleterious nature of mutations,

"The process of mutation is the only known source of the raw materials of genetic variability, and hence of evolution.... the mutants which arise are, with rare exceptions, deleterious to their carriers, at least in the environments which the species normally encounters."[43]

The geneticist, C.P. Martin repeatedly asserts the challenges to the idea that mutations can in anyway be helpful in producing an improved species,

"Accordingly, mutations are more than just sudden changes in heredity; they also affect viability, and, to the best of our knowledge, invariably affect it adversely."[44]

"Mutation does produce hereditary changes, but the mass of evidence shows that all, or almost all, known mutations are unmistakably pathological and the few remaining ones are highly suspect."[45]

"All mutations seem to be in the nature of injuries that, to some extent, impair the fertility and viability of the affected organisms. I doubt if among the many thousands of known mutant types one can be found which is superior to the wild type in its normal environment, only very few can be named which are superior to the wild type in a strange environment."[46]

In the Introduction to Darwin's *The Origin of Species* (6th edition), W.R. Thompson (F.R.S.), Director of the Commonwealth Institute of Biological Control at Ottawa, Canada, is emphatic with regard to the negative aspect of mutations,

> "If we say that it is only by chance that they are useful, we are still speaking too leniently. In general, they are useless, detrimental, or lethal."[47]

Whether they are lethal or just harmful, they are definitely a problem and not at all beneficial, as A.M. Winchester records in his book, *Genetics*,

> "Lethal mutations outnumber visibles by about 20 to 1. Mutations that have small harmful effects, the detrimental mutations, are even more frequent than the lethal ones."[48]

And what is the real common denominator of mutations?

> "The one **systematic** effect of mutation seems to be a tendency towards degeneration."[49] [emphasis in original]

5. Time works against mutations as a mechanism of evolution.

Evolutionists have for a very long time promoted the idea that immense periods of time make it possible for amazing changes to take place in organisms, which we do not see happening in the present world. A good example of this faulty supposition is expressed by George Wald,

> "Time is in fact the hero of the plot. Given so much time the impossible becomes possible, the possible probable and the probable virtually certain. One has only to wait: time itself performs the miracles."[50]

However, time is not on the side of evolution with regard to the use of mutations in effecting positive changes. This concept is here introduced by, J.F. McDonald, in his article entitled, "The Molecular Basis of Adaptation,"

> "[Mutation] is the only apparent source of the major regulatory changes seemingly associated with major adaptive shifts. One possible objection to this view is that the waiting time for such 'hopeful mutations' may be prohibitively long. Moreover, the probability that the occurrence of such a mutation will coincide with a major environment shift requiring an appropriate adaptive response seems infinitesimally small."[51]

Further, as Anna Pai states in her book, *Foundations of Genetics*, research shows that mutations occurring over long periods of time actually weaken the species,

> "This genetic load tends to drag down the genetic quality of the species."[52]

This is in agreement with the following statement by Christopher J. Wills of the University of San Diego,

> "The large majority of mutations, however, are harmful or even lethal to the individual in whom they are expressed. Such mutations can be regarded as introducing a "load," or genetic burden, into the pool."[53]

In his article in Science magazine, "Biological Mechanisms Underlying the Aging Process," Howard J. Curtis indicates the conflict between mutations and time,

> "Certainly the vast majority of mutations must be deleterious, so if the organs of older animals contain appreciable numbers of cells which are carrying mutations, it is a virtual certainty that the organs are functioning less efficiently than they otherwise would."[54]

6. Mutations are inadequate

H. Graham Cannon, Professor of Zoology at Sheffield University, points out the real challenge to evolution with regard to mutations,

> "No new experiment has produced progeny that show entirely new characters, that show entirely new functioning organs. And yet it is

the appearance of new characters in organisms which marks the boundaries of the major steps in the evolutionary scales."[55]

The late Professor Steven Jay Gould of Harvard is very emphatic,

"A mutation doesn't produce major new raw material. You don't make a new species by mutating the species.... That's a common idea people have; that evolution is due to random mutations. A mutation is not the cause of evolutionary change."[56]
Pierre P. Grasse reiterates the same fact, that mutations are,

"...merely hereditary fluctuations around a medium position; a swing to the right, a swing to the left, but no final evolutionary effect....they modify what pre-exists. No matter how numerous they may be, mutations do not produce any kind of evolution."[57]

In the *Encyclopedia of the Life Sciences* we read,

"If mutation were a variation of value to the species, then the evolution of drosophila should have proceeded with extreme rapidity. Yet the facts entirely contradict the validity of this theoretical deduction; for we have seen that Drosophila type has been known since the beginning of the Tertiary period, that is for about fifty million years, and it has not been modified in any way during that time."[58]

Mutations that are easily observed in nature, or that have undergone innumerable mutational experiments, are shown by Grasse to be inadequate as true mechanisms of evolution,

"The fruit-fly (*Drosophila melanogaster),* the favorite pet insect of the geneticists, whose geographical, biotopical, urban, and rural genotypes are now known inside out, seems not to have changed since the remotest times."[59]

Again, Dr. Gould agrees that mutations are inadequate to explain the theorized evolutionary process. In his article, he refers to very minor changes within a species and then says,

"Orthodox neo-Darwinians extrapolate these even and continuous changes to the most profound structural transitions.....Macroevolution (major structural transitions) is nothing more than microevolution (flies in bottles) extended."[60]

However, what we really observe is microevolution (minor changes within the same species, i.e. "flies in bottles," most of which are detrimental to the species, or in this case, the flies). What we do not observe is any real extension of one species into another. The former is observable, the second (extension) is not. The one is science, the latter is imagination.

In challenging the traditional neo-Darwinian theory, Gould asked this question,

"How can such processes change a gnat or a rhinoceros into something fundamentally different?"[61]

Answering his own question in a later article, Gould simply says,

"That theory [orthodox neodarwinian extrapolationalism], as a general proposition, is effectively dead, despite its persistence as textbook orthodoxy."[62]

Further, Grasse explains how mutations that are less easily observed, such as in bacteria, have proven to be equally inadequate,

"Bacteria, the study of which has formed a great part of the foundation of genetics and molecular biology, are the organisms which, because of their huge numbers, produce the most mutants....bacteria, despite their great production of intraspecific varieties, exhibit a great fidelity to their species. The bacillus *Echerichia coli,* whose mutants have been studied very carefully, is the best example. The reader will agree that it is surprising, to say the least, to want to prove evolution and to discover its mechanisms and then to choose as a material for this study a being which practically stabilized a billion years ago.[63]

Science is what you see, not what you imagine. The idea that genetic changes can occur by means of mutations, which over long periods of time can produce a higher and more complex organism, is indeed a theory. However, for all of the hype made regarding this theory, years of research has shown that this concept is totally devoid of observational support. As, Michael Pitman explains,

> "Do we, therefore, ever see mutations going about the business of producing new structures for selection to work on? No nascent organ has ever been observed emerging, though their origin in pre-functional form is basic to evolutionary theory. Some should be visible today, occurring in organisms at various stages up to integration of a functional new system, but we don't see them: there is no sign at all of this kind of radical novelty. Neither observation nor controlled experiment has shown natural selection manipulating mutations so as to produce a new gene, hormone, enzyme system or organ."[64]

Many researchers have written about the amazing complexity of the genes, and the information they have provided certainly indicates that such a unique and complex organ could not ever come into existence by chance. Frank B. Salisbury of Utah State University sums it up this way,

> "If life really depends on each gene being as unique as it appears to be, then it is too unique to come into being by chance mutations."[65]

Jeffrey S. Wicken, Professor of Biochemistry at Pennsylvania State University, Behrend College, in his article about complexity in evolution, concludes,

> "As a generative principle, providing the raw material for natural selection, random mutation is inadequate both in scope and theoretical grounding."[66]

In the final analysis, the theory of evolution is maintained, not because of the observable facts of science, but in spite of them, as these comments by Muller illustrate,

> "It is entirely in line with the accidental nature of mutation that extensive tests have agreed in showing the vast majority of them detrimental to the organism in its job of surviving and reproducing, just

as changes accidentally introduced into any artificial mechanism are predominantly harmful to its useful operation. ...

It is nevertheless to be inferred that all the superbly interadapted genes of any present-day organism arose through just this process of accidental natural mutation."[67]

Finally, Lynn Margulis, referring to evolutionary changes, forthrightly lays out the obvious,

"I have seen no evidence whatsoever that these changes can occur through the accumulation of gradual mutations."[68]

As a result of his scientific research, Nils Heribert-Nilsson, the Swedish geneticist concluded,

"It is, therefore, absolutely impossible to build a current evolution on mutations or on recombinations."[69]

7. How about super mutations or a super fast mutation rate?

Many years ago, Goldschmidt came up with the idea that evolution proceeded in giant jumps. One day a reptile laid an egg and a bird hatched out of the egg. This was referred to as the "hopeful monster" theory. It was, at first, criticized by many scientists, but eventually it was revisited and revised by Gould and Lewontin to make it more palatable. In their opinion it didn't happen overnight but in a very short period of time geologically. Their theory was called "punctuated equilibriunism". Ernst Mayr has voiced quite negative comments with regard to the "hopeful monster" idea,

"The occurrence of genetic monstrosities by mutation ... is well substantiated, but they are such evident freaks that these monsters can be designated only as 'hopeless.' They are so utterly unbalanced that they would not have the slightest chance of escaping elimination through stabilizing selection .... the more drastically a mutation affects the phenotype, the more likely it is to reduce fitness. To believe that such a drastic mutation would produce a viable new type, capable of occupying a new adaptive zone, is equivalent to believing in miracles .... The finding of a suitable mate for the 'hopeless monster' and the establishment of reproductive isolation from the normal members of the parental population seem to me insurmountable difficulties."[70]

Chapter Three End Notes

[1] Ernst Mayr quoted in *Time Frames: The Rethinking of Darwinian Evolution and the Theory of Punctuated Equilibria*, by Niles Edredge, Simon & Schuster, 1985, p. 33
[2] The Human Genome Project, Announcement from the University of Texas Southwestern Medical School, May 6, 1993
[3] *Evolution: Change Over Time*, Prentice Hall, 1993 p 49
[4] Bernice Essenfeld, Carol Gontang, Randy Moore, *Biology*, Addison-Wesley, 1994, p. 250
[5] *Ibid.*
[6] *Ibid.*
[7] Alton Biggs, Chris Kapicka, Linda Lundgren, *Biology, The Dynamics of Life*, Glencoe, McMillian/McGraw-Hill, 1995, p. 432
[8] *Ibid.*, p. 437
[9] Sylvia S. Mader, *Biology,* McGraw Hill, 2001, p.304
[10] Ernst Mayr, *Populations, Species and Evolution: An Abridgment of Animal Species and Evolution,* Belknap Press, 1970, p. 102
[11] Francisco J. Ayala, "The Mechanisms of Evolution," *Scientific American*, 1978, p. 63
[12] Francisco J. Ayala, "Teleological Explanations in Evolutionary Biology," in Philosophy of Science, March 1970, p. 3
[13] Richard Dawkins, "Creation and Natural Selection," *New Scientist*, vol. 111, September 1986, p. 38
[14] D. Erwin & J. Valentine, "'Hopeful Monsters,' Transpsons, and Metazoan Radiation," in *Proceedings National Academy of Sciences*, Vol. 81, 1984, p. 5482
[15] John Endler & T. McLellan, "The Process of Evolution: Toward a Newer Synthesis," *Annual Review of Ecology and Systematics, 19*, 1988, p. 397
[16] *World Book Encyclopedia*, 1966 Edition
[17] Francisco Ayala, "The Mechanisms of Evolution," *Scientific American*, vol. 239, September 1978, p. 64
[18] *Ibid.*
[19] Environmental Mutagenic Society, "Environmental Mutagenic Hazards," *Science,* Vol. 187, February 14, 1975, p. 504
[20] P.P. Grasse, *Evolution of Living Organisms*, Academic Press, 1977, pp. 97-98
[21] *Ibid.*
[22] *Ibid.*, p. 103
[23] Francisco Ayala, "The Mechanisms of Evolution," *Scientific American*, vol. 239, September 1978, p. 59
[24] H.J. Muller, "Radiation Damage to the Genetic Material," *American Scientist*, Vol. 38, Jan. 1950, p. 35
[25] Julian Huxley, *Evolution in Action*, 1953, (2007 edition published by Harper & Row) p. 37
[26] Theodosius Dobzhansky, *Evolution, Genetics and Man*, New York: John Wiley & Sons, 1955, p. 150
[27] James F. Crow, "Genetic Effects of Radiation," *Bulletin of the Atomic Scientists,* Vol. 14, Jan. 1958, pp. 19-20
[28] *Ibid.*
[29] George Wald, as quoted by Murray Eden, "Inadequacies of Neo-Darwinian Evolution as a Scientific Theory," *Mathematical Challenges to the Neo-Darwinian Interpretation of Evolution*, editors Paul S. Moorhead and Martin M. Kaplan, pp. 18-19
[30] Environmental Mutagenic Society, *op cit*, p. 503.
[31] *Ibid,* p. 512
[32] C.P. Martin, "A Non-Geneticist Looks at Evolution," *American Scientist,* vol. 41, January 1953, p. 101
[33] *Ibid,* p. 103

[34] Paul F. Brandwin, Robert Stollberg, Arthur Greenstone, Warren E. Yasso, & Daniel J. Brovey, *Life Its Forms and Changes,* Second Edition, Harcourt Brace Jovanovich, Inc., 1972, p. 479
[35] *Ibid.*, p. 480
[36] *Ibid.*
[37] *Ibid.*
[38] *Ibid.*
[39] *Ibid.*
[40] Francisco J. Ayala, "Genotype, Environment and Population Numbers," *Science,* Vol. 162. Dec. 27, 1968, p. 1436
[41] James F. Crow, "Genetic Effects of Radiation," *Bulletin of the Atomic Scientists,* Vol. 14, January 1958, pp. 19-20
[42] Warren Weaver et al., "Genetic Effects of Atomic Radiation," Science, vol. 123, June 29, 1956, p. 1159
[43] Theodosius Dobzhansky, "On Methods of Evolutionary Biology and Anthropology," *American Scientist,* Winter, December 1957, p. 385
[44] C.P. Martin, "A Non-Geneticist Looks at Evolution," American Scientist, January 1953, p. 102
[45] *Ibid.*, p. 103
[46] *Ibid.*, p. 100
[47] W.R. Thompson, "Introduction to *The Origin of Species,*" Everyman Library No. 811, E.P. Dutton & Sons, 1956; reprint edition: J.M. Dent and Sons, Ltd., 1967, p. 10
[48] A.M. Winchester, *Genetics,* 5th edition, Houghton Mifflin Co., 1977, p. 356
[49] Sewall Wright, "The Statistical Consequences of Mendelian Heredity in Relation to Speciation," *The New Systematics,* editor Julian Huxley, Oxford University press, 1949, p. 174
[50] George Wald, "The Origin of Life," *Scientific American,* August 1954, p. 48
[51] J.F. McDonald, "The Molecular Basis of Adaptation," in *Annual Review Ecology and Systematics,* 14, 1983, p. 94
[52] Anna Pai, *Foundations of Genetics,* McGraw-Hill, 1974
[53] Christopher Wills, "Genetic Load," *Scientific American,* March 1970, p. 98
[54] Howard J. Curtis, "Biological Mechanisms Underlying the Aging Process," *Science,* Vol. 141, August 23, 1963, p. 688
[55] H. Graham Cannon, *The Evolution of Living Things,* Manchester University Press, 1958 p. 82
[56] Stephen J. Gould, Lecture at Hobart and William Smith College, Feb. 14, 1980
[57] P.P. Grasse, *Evolution of Living Organisms,* Academic Press, 1977, p. 88
[58] Michel Delsol, *Encyclopedia of the Life Sciences,* vol. 2, p. 34
[59] Pierre P. Grasse, *Op cit,* p. 130
[60] Stephen J. Gould, "Return of the Hopeful Monster," *Natural History,* June/July 1977
[61] *Ibid.*
[62] Stephen J. Gould, "Is a New and General Theory of Evolution Emerging?", *Paleobiology,* 1980, p. 120
[63] Pierre P. Grasse, *Evolution of Living Organisms,* Academic Press, 1977, p. 87
[64] Michael Pitman, *Adam and Evlution,* Rider, 1984, pp. 67-68
[65] Frank B. Salisbury, "Natural Selection and the Complexity of the Gene," Nature, Vol. 224, 25 October 1969, p. 342
[66] Jeffrey S. Wicken, "The Generation of Complexity in Evolution: A Thermodynamic and Information-theoretical Discussion," *Journal of Theoretical Biology,* Vol. 77, April 1979, pp. 351-352
[67] H.J. Muller, "How Radiation Changes the Genetic Constitution," *Bulletin of the Atomic Scientists,* Vol. 11, No. 9, November 1955, p. 331
[68] Lynn Margulis, as quoted by Charles Mann, "Lynn Margulis: Science's Unruly Earth Mother," *Science,* Vol. 252, 19 April 1991, p. 379
[69] N. Heribert Nilsson, *Synthetische Artbildung,* Verlag CWK Gleerup, 1953, p. 1157
[70] Ernst Mayr, *Populations, Species, and Evolution,* Belknap Press, 1970, p. 235

NATURAL SELECTION

Introduction

Long before Charles Darwin wrote *Origin of Species* many scientific minded people had been experimenting with various plants and animals using a process called artificial selection. As time went by, they acquired more knowledge, experience, and had more success in their attempts to develop variations within certain species of animals or plants. Charles Darwin, a naturalist, was well aware of the phenomenon of artificial selection and of what could be achieved by its application to the world of nature. In the end, this concept became a primary tool in his effort to become a prominent figure in the area of biology. During this era in English history most of the people believed that God, the Creator, had brought into existence the various kinds of animals by the Word of His creative power. This was referred to as "special creation." One of the strongest arguments espoused by the academics in favor of this concept was that of intelligent design. They argued that since we know that our clocks, street lamps, and kitchen utensils owed their existence, not to chance, but to intelligent design, it therefore stood to reason that the magnificent designs found in amazing creatures, such as butterflies, tropical fish, and elephants, owed their existence to an intelligent designer as well. The fact that the book of Genesis stated that this was the case further amplified the soundness of this argument. However, by the mid 1800s certain influential people in England wished to see the origin of the various species as occurring by an entirely naturalistic means. Darwin was a man who was willing to provide a solution to this challenge. Thus, it became his burden to endeavor to explain intelligent design without having to refer to an intelligent designer. In order to achieve this goal Darwin adopted the idea that in the same way artificial selection was used to obtain variant breeds of dogs in a short period of time, natural selection could, over millions of years of time, make it possible for a simple species to evolve into a more complex one. From that time to the present, the concept of natural selection has become a standard "mechanism of evolution," and is featured prominently in the State adopted Biology textbooks. As a consequence, the academic world has dismissed the Genesis account of creation (creation by the Word of God), calling it irrelevant. In light of the preceding remarks, it should come as no surprise to find that the theory of evolution has become "scientific" support for the ideologies of materialism and atheism.

Textbook Evolution

"The heart of Darwin's theory of evolution is its assertion that natural selection is the mechanism responsible for evolution."[1] (George B. Johnson & Peter H. Raven, *Biology, Principles and Explorations*, Holt, Rinehart & Winston, Inc.)

"In response to their environments species evolve, or change over time. The changes that make organisms better suited to their environments develop through a process called natural selection."[2] (___, *Science Explorer, Focus on Earth Science*, Prentice Hall)

"As a result, species today look different from their ancestors. Each living species has descended, with changes, from their ancestors."[3] (K.R. Miller and J. Levine, *Biology,* Prentice Hall)

"Natural selection is a primary process that brings about evolution by selecting which individuals will survive, reproduce, and pass on their genes to the next generation."[4] (Eldon Enger, Frederick Ross, & David Bailey, Concepts in Biology, 13th edition, McGraw-Hill)

"Natural selection results in the evolution of offspring that are better adapted to that environment. If their environments differ enough, separate populations of the same species can become very dissimilar. Over time, populations of the same species that differ genetically because of adaptations to different living conditions become what biologists call subspecies."[5] (*Ibid.*, p.291)

Moths

Textbook Evolution

"The evolution of camouflage in moths and the development of physiological resistance to chemicals in organisms provide examples of the process of evolution."[6] (Alton Biggs, Chris Kapicka, Linda Lundgren, *Biology, The Dynamics of Life*, Glencoe,McMillian/McGraw-Hill)

"Although the dark variety of Biston betularia is not a new species, if enough 'microchanges' accumulate in a population and the population becomes reproductively isolated, microevolution can produce a new species."[7] (Bernice Essenfeld, Carol Gontang, Randy Moore, *Biology,* Addison-Wesley)

In the Holt 1994 text, *Biology Visualizing Life*, there is at Table 9.2 the following information which is entitled "How Kettlewell Demonstrated Natural Selection in Peppered Moths." Pictures of both the light and dark moths are provided for the students to observe. Using the lower half of page 188, the students learn in three steps what Kettlewell did and what Kettlewell found.

"Step 1. Kettlewell knew that coloration was an inherited trait. He raised large numbers of both light and dark moths in the laboratory. He released equal numbers of light and dark moths into a forest near Birmingham, England, where trees were blackened by soot. For later recognition, each of these moths was marked with a dot on the underside of its wings."

"Kettlewell set out rings of traps to recapture moths that survived. He found that two-thirds of the recaptured moths were dark. More moths that matched the dark tree trunks had survived."

"Step 2. Again using marked moths, Kettlewell released equal numbers of light and dark moths into an unpolluted forest in Dorset, England.Trees here were white, not black."

"When Kettlewell set out traps, two-thirds of the moths that were recaptured were light. Again, more moths that matched the color of the tree trunks survived."[8] (George B. Johnson, *Biology Visualizing Life,* Holt, Rinehart, & Winston)

It will be useful for the reader, as we review the evidence to the contrary, to remember this grand statement shown below which refers to the moth experience listed above.

"... the most striking evolutionary change ever witnessed by man."[9] (*International Wildlife Encyclopedia*, Burton & Burton)

"Consider the English peppered moths, which inhabit forested areas. Before the Industrial Revolution, most peppered moths were a lighter color. Tree trunks in rural England were covered with light gray lichens, allowing light-colored moths to blend perfectly with the tree trunks. Darker moths were seen occasionally, probably due to mutations. But darker moths were easy for predators to spot, and they were quickly eaten by the birds."[10] (*Biology, The Web of Life*, Second Edition, Scott-Foresman-Addison Wesley)

"The moths that blend in with their background are said to be camouflaged."[11] (*Ibid.*)

"Before the Industrial Revolution, soot was rare in the English countryside. A light-colored moth was difficult to see against the clean tree bark. After several years...the bark was darkened by the soot of burning coal, a light-colored moth stood out against the darkened tree bark."[12] (*Ibid.*)

"Kettlewell learned how to capture both light and dark-colored forms of the peppered moth and then managed to raise them in captivity. He also learned to mark living moths in such a way that birds could not see the marks. Kettlewell then released equal numbers of light and dark-colored moths in two types of areas. In one area, trees were normally colored. In the other area, they were blackened by soot. Later on, he recaptured, sorted and counted all the marked moths he could. What type of results do you think Kettlewell needed to either prove or disprove the hypothesis?"[13] (*Ibid.*)

Both the 1991 and 1998 Miller-Levine Biology text say the exact same words with regard to summarizing Kettlewell's experiment,

"Kettlewell found that in unpolluted areas, more of his light-colored moths had survived. In soot-blackened areas, more of his dark-colored moths had survived. Thus Kettlewell showed that in each environment the moths that were better camouflaged had the higher survival rate. It was logical to conclude that when soot darkened the tree trunks in an area, natural selection caused the dark-colored moths to become more common. Today Kettlewell's work is considered to be a classic demonstration of natural selection in action."[14] (Kenneth R. Miller and Joseph Levine, *Biology*, Prentice Hall)

It is important at this point to reach back to the Holt 1994 text, *Biology Visualizing Life*, which at Table 9.2 includes this statement regarding Kettlewell's elaborate experiment,

"Step 3. Kettlewell set up hidden cameras in both forests to record the capture of moths by birds."

"Films showed that birds were more likely to capture light moths on the dark trunks near Birmingham. In Dorset, birds were more likely to eat the dark moths, which were obvious against the light trunks."[15] (George B. Johnson, *Biology Visualizing Life*, Holt, Rinehart and Winston)

Science Versus

After reading these many quotes from the textbooks, it should be clear to the reader that the peppered moth experiments by Kettlewell have occupied a large position in their role as a major support for the theory of evolution. However, as will soon be demonstrated, this was all pure fabrication. First, it must be stated that for a very long time top evolutionary researchers, such as the British Zoologist, L. Harrison Matthews, have known that, while the peppered moth experiment did demonstrate natural selection in action, it did not demonstrate evolution in action.

"Although (peppered moth) experiments beautifully demonstrate natural selection, or survival of the fittest, they do not show evolution in progress, for however the populations may alter...all the moths remain from beginning to end, Biston betularia."[16]

So, from the very beginning the peppered moth experiments may have been useful in showing natural selection, but were not useful in showing evolution. More important research on the moth scenario has been performed since the days of Kettlewell and the results of these experiments only serve to underscore the kind of pseudo science that has been long used to support the theory of evolution and make it believable to the general public.

Consider for instance, the statements made by Miller and Levine in the high school text, *Biology*, regarding this business about the moths Kettlewell filmed while they were resting on trees so that he could observe and record how the birds ate them,

> "England's peppered moth provides an example of natural selection in action. It also offers us a chance to study the sorts of experiments that can be used to test evolutionary theory. The story is as follows. The peppered moth spends much of the daytime resting on the bark of oak trees."[17]

Over time, other people tried to duplicate Kettlewells' experiments. They were not successful. Why they were not successful is discussed in the book, *Melanism: Evolution in Action* by Michael Majerus. This book was reviewed by Jerry Coyne of *Nature* Magazine,

> "Majerus notes that the most serious problem is that *B. betularia* probably do not rest on tree trunks -- exactly two moths have been seen in such a position in more than 40 years of intensive search. The natural resting spots are, in fact, a mystery."[18]

What we have here in the above statements from the textbooks are not accurate scientific information, but simply propaganda. Coyne reports,

> "This alone invalidates Kettlewell's release-recapture experiments, as moths were released by placing them directly onto tree trunks, where they are highly visible to bird predators."[19]

> "Kettlewell also released his moths during the day, while they normally choose resting places at night."[20]

> "The story is further eroded by noting that the resurgence of *typica* occurred well before lichens recolonized the polluted trees, and that a parallel increase and decrease of the melanic form also occurred in industrial areas of the United States, where there was no change in the abundance of the lichens that supposedly play such an important role."[21]

"Unfortunately, most of the work described is inconclusive; despite the widespread occurrence of melanism, its evolutionary significance is nearly always unknown."[22]

"Finally, the results of Kettlewell's behavioral experiments were not replicated in later studies: moths have no tendency to choose matching backgrounds."[23]

Majerus, himself, goes on to report,

"Second, during the course of the more recent work, it has become obvious that much of both the experimental and theoretical work suffers from artificiality. In many experiments, moths are placed into artificial situations, which may affect their own behavior or that of their predators. For example, in most predation experiments peppered moths have been positioned on vertical tree trunks, despite the fact that they rarely chose such surfaces to rest upon in the wild."[24]

"When Kettlewell released peppered moths in Birmingham and Dorset to study bird predation, he did so by placing live moths onto tree trunks where they could be observed (Kettlewell 1955a, 1956)."[25]

"Most subsequent experiments to determine predation rates on peppered moth morphs have used dead moths, glued in 'natural postures' to tree trunks."[26]

"Data on the natural resting sites of the peppered moth are pitifully scarce, and this in itself suggests that peppered moths do not habitually rest in exposed positions on tree trunks. Many nocturnal moths do rest by day on tree trunks and searching trunks has long been a recognized method of collecting by lepidopterists. However, the number of published records of peppered moths being found on tree trunks is negligible. This is emphasized in an admission by Sir Cyril Clarke [Clarke et al. 1985]: 'all we have observed is where the moths do not spend the day. In 25 years we have only found two *betularia* on the tree trunks or walls adjacent to our traps [one on an appropriate background and one not], and none elsewhere.'"[27]

Further, Professor Kauri Mikkola, who is associated with the Finnish Museum of Natural History, lets us know that,

> "The normal resting place of the Peppered Moth is beneath small, more or less horizontal branches (but not on narrow twigs), probably high up in the canopies"[28]

Professor Cyril A. Clarke, physician and geneticist, previously with the University of Liverpool in England, confirms that,

> ".... exposed tree trunks are precisely where the moths do NOT spend the day"[29]

Finally, with regard to moths, the coloration and attempts to show natural selection in process here has continually come short of explaining the real challenge to evolution--the origin of moths,

> "It is now approximately half a century since the neo-Darwinian synthesis was formulated. A great deal of research has been carried on within the paradigm it defines. Yet the successes of the theory are limited to the minutiae of evolution, such as the adaptive change in coloration of moths; while it has remarkably little to say on the questions which interest us most, such as how there came to be moths in the first place."[30]

From the creation standpoint it has long been argued that natural selection can play a significant role in screening out unfit or less viable life forms. In this way it preserves the healthy and vibrant life forms of any particular species, but it does not improve those particular life forms, let alone create any new life forms. A variety of evolutionary researchers, such as the late Stephen J. Gould of Harvard, have come to acknowledge this unique screening roll of natural selection as well as its limitations.

> "The essence of Darwinism lies in a single phrase: natural selection is the creative force of evolutionary change. No one denies that natural selection will play a negative role in eliminating the unfit. Darwinian theories require that it create the fit as well."[31]

In this connection Niles Eldredge, curator of the Museum of Natural History, also states,

"But natural selection per se does not work to create new species."[32]

Miller and Levine in their textbook, *Biology,* say that the evolutionary process (descent with modification) produce,

"A single 'tree of life' that links all living things on earth."[33]

This is an idealistic view of the whole evolutionary story. It is the party line from a long time ago, but it is a bland statement, which, when subjected to scrutiny, is not supported by the evidence from fossils or from what we know about how natural selection actually works. As American paleontologist Steven Stanley of the University of Hawaii at Manoa explains,

"If most evolutionary changes occur during speciation events and if speciation events are largely random, natural selection, long viewed as the process guiding evolutionary change, cannot play a significant role in determining the overall course of evolution."[34]

He further comments,

"The reductionist's view that evolution can ultimately be understood in terms of genetics and molecular biology is clearly in error"[35]

Lewontin of Harvard University also reports,

"...in other words, natural selection over the long run does not seem to improve a species' chance of survival but simply enables it to "track", or keep up with the constantly changing environment."[36]

Some Biologists have also come to recognize and acknowledge that the concepts of natural selection as used as an evolutionary mechanism is not a real explanation of anything, but simply a tautology. By tautology, they mean the expression of a concept that, in attempting to prove a point, they simply repeat the concept with different words. Professor Gregory Alan Pesely, clarifies this erroneous process in his article, "The Epistemological Status of Natural Selection,"

"One of the most frequent objections against the theory of natural selection is that it is a sophisticated tautology. Most evolutionary biologists seem unconcerned about the charge and only make a token effort to explain the tautology away. The remainder, such as Professors Waddington and Simpson, will simply concede the fact. For them, natural selection is a tautology which states a heretofore unrecognized relation: The fittest -- defined as those who will leave the the most offspring -- will leave the most offspring. What is most unsettling is that some evolutionary biologists have no qualms about proposing tautologies as explanations. One would immediately reject any lexicographer who tried to define a word by the same word, or a thinker who merely restated his proposition, or any other instance of gross redundancy; yet no one seems scandalized that men of science should be satisfied with a major principle which is no more than a tautology."[37]

Arthur Koestler presents the tautology in a fairy tale setting,

"Once upon a time, it all looked so simple. Nature rewarded the fit with the carrot of survival and punished the unfit with the stick of extinction. The trouble only started when it came to defining fitness....Thus natural selection looks after the survival and reproduction of the fittest, and the fittest are those which have the highest rate of reproduction, ...we are caught in a circular argument which completely begs the question of what makes evolution evolve."[38]

As C. Loring Brace, Curator of Biological Anthropology and Professor of Anthropology at the University of Michigan, reveals, natural selection is waning in its importance to the theory of evolution,

"Readers of *American Scientist* may not realize the extent to which a major part of the field of biology and almost all of paleontology has rejected Darwin's insights concerning organic evolution. Natural selection is dismissed as contributing nothing more than "fine-tuning" and adaptation is largely ignored in practice."[39]

Resistance To Antibiotics

Textbook Evolution

> "Figure 18.6 - Examples of bacterial resistance to antibiotics, such as penicillin, provide scientists with direct evidence for evolution in progress."[40] (Alton Biggs et al., *Biology: The Dynamics of Life*, Glencoe,McMillian/McGraw-Hill)

> "The development of physiological resistance in species of bacteria, insects, and plants is direct evidence of evolution."[41] (Alton Biggs et al., *Biology: The Dynamics of Life*, Glencoe/McGraw-Hill)

In the 2004 Glencoe text, at the top of page 399, they have three factual statements:

1. "The bacteria in a population vary in their ability to resist antibiotics."

2. "When the population is exposed to an antibiotic, only the resistant bacteria survive."

3. "The resistant bacteria live and produce more resistant bacteria."

Then on the same page under the title "Figure 15.5" they introduce the students to an assumption which is devoid of factual support,

> "The development of bacterial resistance to antibiotics is direct evidence for evolution."[42] (Alton Biggs, et al., *Biology: The Dynamics of Life*, Glencoe)

> "The development of physiological resistance in species of bacteria, insects, and plants is direct evidence of evolution."[43] (Alton Biggs et al., *Biology: The Dynamics of Life*, Glencoe Science, *National Geographic*, McGraw-Hill/Glencoe)

> "Figure 14-21 traces how pesticide resistance evolves. Most survivors of the first pesticide treatments were insects with genes that somehow enabled them to resist the chemical attack. Their offspring inherited the genes for pesticide resistance. In each generation, the percentage of

pesticide-resistant individuals in the beetle population increased. The population underwent evolutionary change that resulted in adaptation to a change in the chemical environment--the presence of the pesticide."[44] (Neil A. Campbell, Brad Williamson, & Robin J. Heyden, *Biology: Exploring Life*, Pearson)

Science Versus

On page 245 of the college text, *Essential Biology with Physiology*, **Figure 13.1**, is entitled, **Evolution of pesticide insect populations**. The word "evolution" is very misleading, because when you study the information regarding pesticide-resistant organisms, it is clearly evident that no evolution has taken place at all. For example, Francisco Ayala, past president and Chairman of the Board of the American Association for the Advancement of Science reports that:

> "Insect resistance to pesticide was first reported in 1947 for the housefly (Musca domestica) with respect to DDT. Since then the resistance to pesticides has been reported in at least 225 species of insects and other arthropods."[45]

Evolution of resistance did not take place for, as Ayala acknowledges,

> "The genetic variants required for resistance to the most diverse kinds of pesticides were apparently present in every one of these populations exposed to these man-made compounds."[46]

Further, in an article in *Reuters* entitled, "Wild Animals Resistant to Antibiotics," we are informed that,

> "Doctors and health officials are becoming increasingly concerned about the rise of antibiotic-resistant bacteria."

> "U.S. doctors estimate as many as 70 percent of the bacteria in infections that people get in hospitals now are resistant to at least one antibiotic."[47]

In fact, the aforementioned college text correctly relates the following:

"A relatively small amount of the poison dusted onto a crop may kill 99% of the insects. However, the relatively few survivors of the first pesticide wave are insects with genes that somehow enable them to resist the chemical attack (**Figure 13.1**). For example, the lucky few may carry gene coding for enzymes that destroy the pesticide. The poison kills most members of the insect population, leaving only the resistant individuals to reproduce. And when they do, their offspring inherit the genes for pesticide resistance. In each generation, the proportion of pesticide-resistant individuals in the insect population increases, making subsequent spraying less and less effective."[48]

The insects that survive the pesticide and those that did not were the same exact species. Before, after, and during this experience, the species remained exactly the same. No evolution from one species to another has taken place. It is misleading to say so. The same text used above also tells the students that, "The result of natural selection is evolutionary adaptation..."[49] As one can clearly see from the insecticide-resistant experience, adaptation is not the result of natural selection, natural selection is the result of a lack of adaptation. The insects whose genes made it possible for them to be protected from the harmful effects of the pesticide survived. Those insects whose genes were unable to protect them from the harmful effect of pesticide were "selected" out. They did not survive. In such cases as this, natural selection preserves the species that are already adapted and eliminates those that are not.

Bacteria

In addition to telling the students that various insects have evolved a resistance to DDT, they also tell the students that certain bacteria have evolved a resistance to modern antibiotics.

However, the real explanation for the bacteria resisting antibiotics is essentially the same as that for the insects which were resistant to various insecticides. A small percentage of the population is already resistant, and when non-resistant bacteria is eliminated, the resistant bacteria multiply. The fact that some bacteria have a native resistance to certain antibiotics has been graphically demonstrated by the discovery of frozen human bodies in the Arctic area. They perished prior to the time that antibiotics were invented and were discovered long after the antibiotics had been in use. Nevertheless, the bacteria in these bodies has been found to be resistant to modern antibiotics. This information was revealed in a Medical Tribune article in 1988,

> "Current theories suggest that antibiotic resistance is linked to long-term exposure to antibiotics. Needless to say, antibiotics were not developed until long after these 19th-century bacteria and their hosts had been buried in Arctic permafrost."[50]

> "It may be time to rethink our thoughts about the mechanisms for antibiotic resistance patterns. ...the anaerobic bacteria, from the bowels of three members of an 1845 Arctic expedition, have survived 140 years and are showing resistance patterns to modern antibiotics."[51]

Anthony Hart of the University of Liverpool (referenced above in the article from *Reuters*), states,

> "Our results show that resistance to antibiotics is widespread in at least some wild populations, even though these have never to our knowledge been exposed to antibiotics, and they undermine the presumption that resistance will decline in the absence of antibiotic treatment."[52]

These results further undermine the fiction that bacteria evolve resistance to antibiotics.

Microevolution does not Lead to Macroevolution

Microevolution is simply variation within a species. From the creation standpoint this indicates that the Creator created within each original animal kind the capacity for variation so that, over long periods of time with changing environmental conditions, an adequate number of organisms could survive. Microevolution is observed everywhere, but macroevolution is observed nowhere. The Addison-Wesley textbook, *Biology*, attempts to lead the students to believe that microevolution does lead to macroevolution,

"Although the dark variety of Biston betularia is not a new species, if enough 'microchanges' accumulate in a population and the population becomes reproductively isolated, microevolution can produce a new species."[53]

Unfortunately, the students are not informed of all the incoming information regarding this subject. For instance, a large conference was held in Chicago for the purpose of examining the possibilities that this might indeed be the case. Roger Lewin's review in *Science* of the many papers presented there does not encourage faith in this concept,

"The central question of the Chicago conference was whether the mechanisms underlying microevolution can be extrapolated to explain the phenomena of macroevolution. At the risk of doing violence to the positions of some of the people at the meeting, the answer can be given as a clear 'No.'"[54]

With regard to obtaining new traits and adaptation, the Holt text, *Modern Biology*, tells the students,

"If a trait both increases the reproductive success of an organism *and* is inherited, then that trait will tend to be passed on to many offspring. A population of organisms adapt to their environment as their proportion of genes for favorable traits increases. The resulting change in the genetic makeup of a population is evolution."[55]

How are these authors proposing that this new trait develops in the first place since acquired traits are not inherited and natural selection has nothing to do with the creation of new traits? It has been made clear by a number of researchers that natural selection by itself is not adequate, for natural selection deals only with the traits that are already present. It is not able to create new traits. In his article on adaptation Lewontin emphasizes this point,

"...evolution can not be described as a process of adaptation because all organisms are already adapted."[56]

He further states,

"The relation between adaptation and natural selection does not go both ways. Whereas greater relative adaptation leads to natural selection, natural selection does not necessarily lead to greater adaptation."[57]

One of his examples has to do with plants adapting to a region that is becoming dry,

"...plants may respond by evolving a deeper root system or a thicker cuticle on the leaves, but only if their gene pool contains genetic variation for root length or for cuticle thickness."[58]

Lewontin explains that,

"...natural selection operates essentially to enable organisms to maintain their state of adaptation rather than to improve it."[59]

He also acknowledges that extinct species may have been just as fit to survive as our modern ones,

"...natural selection over the long run does not seem to improve a species' chance of survival, but simply enables it to "track" or keep up with, the constantly changing environment."[60]

From the standpoint of Creation these statements demonstrate that natural selection works because each created kind was endowed with sufficient variety to multiply and fill the earth in all of its ecological and geographical aspects.

Summary

In conclusion, these summary remarks listed below should, coming after the detailed information written above, enable the reader to see that natural selection does not qualify as a mechanism for evolution. In the first two quotes Steven Stanley discounts natural selection,

> "Gradual evolutionary change by natural selection operates so slowly within established species that it cannot account for the major features of evolution."[61]

> "...natural selection, long viewed as the process guiding evolutionary change, cannot play a significant role in determining the overall course of evolution."[62]

In his article, "Darwin's Mistake" (*Harper's Magazine*), the editor of the Washington Review, Tom Bethell, wrote,

> "It is surprising that so little word of it has leaked out, because it seems to have been one of the most important academic debates of the 1960's, and as I see it the conclusion is pretty staggering: Darwin's theory, I believe, is on the verge of collapse. In his famous book, *On the Origin of Species by Means of Natural Selection, or the Preservation of Favored Races in the Struggle for Life*, Darwin made a mistake sufficiently serious to undermine his theory.
>
> And that mistake has only recently been recognized as such. The machinery of evolution that he supposedly discovered has been challenged, and it is beginning to look as though what he really discovered was nothing more than the Victorian propensity to believe in progress. At one point in his argument, Darwin was misled."[63]

Pierre P. Grasse of the Sorbonne is very emphatic about the inadequacy of natural selection as a mechanism,

> "The role assigned to natural selection in establishing adaptation, while speciously probable, is based on not one single sure datum..."[64]

Stephen Gould informs us that,

"Many evolutionists now doubt exclusive control by selection upon genetic change within local populations. Moreover even if local populations alter as the synthesis maintains, we now doubt that the same style of change controls events at the two major higher levels; speciation and patterns of macroevolution."[65]

In referencing Brooks and Wiley, Roger Lewin, writing in *Science*, states,

"Natural selection, a central feature of neo-Darwinism, is allowed for in Brooks and Wiley's theory, but only as a minor influence. 'It can affect survivorship' says Brooks. 'It can weed out some of the complexity and so slow down the information decay that results in speciation. It may have a stabilizing effect, but it does not promote speciation. It is not a creative force as many people have suggested.'"[66]

This concept has been amplified by a number of writers, as listed below (Lewontin, Goldschmidt, and Rosen), that natural selection does not lead to speciation,

"The relation between adaptation and natural selection does not go both ways. Whereas greater relative adaptation leads to natural selection, natural selection does not necessarily lead to greater adaptation."[67]

"Such an assumption is violently opposed by the majority of geneticists, who claim that the facts found on the subspecific level must apply also to the higher categories. Incessant repetition of the unproved claim, glossing lightly over the difficulties, and the assumption of an arrogant attitude toward those who are not so easily swayed by fashions in science, are considered to afford scientific proof of the doctrine. It is true that nobody thus far has produced a new species or genus, etc., by macromutation. It is equally true that nobody has produced even a species by the selection of micromutations."[68]

"Although natural selection theory fails to explain the origin of evolutionary novelties, its greatest shortcomings in terms of evolutionary theory is that it fails to explain evolutionary diversity."[69]

Dr. Colin Patterson, the senior paleontologist at the British Museum of Natural History, said this about natural selection,

"No one has ever produced a species by mechanisms of natural selection. No one has gotten near it and most of the current argument in neo-Darwinism is about this question...."[70]

The whole idea of evolution is the slow and gradual formation of new species by properties inherent in nature. Yet, as shown above, natural selection, has not been able to bring about a new species. Further, after millions of dollars of research, as Keith S. Thompson contributes in his article in *American Scientist*, the whole idea of speciation remains the "central mystery,"

"Evolution is...troubled from within by the troubling complexities of genetic and developmental mechanisms and new questions about the central mystery--speciation itself."[71]

How exactly, apart from special creation, does a species form? In the evolutionary scenario this, as Stephen R. Palumbi, Senior Fellow of Stanford Woods Institute for the Environment, comments, remains a continuing problem,

"The formation of species has long represented one of the most central, yet also one of the most elusive, subjects in evolutionary biology."[72]

Except in the case of forced mutations by x-ray or some other device, the burden of finding evidence for evolution has been placed on the paleontologists who work with fossils. In the modern world, no one can find any evidence of a new species evolving. This is remarkable, especially in the light of Lewontin's report,

"There are on the order of two million species now living, and since at least 99.9 percent of the species that have ever lived are now extinct, the most conservative guess would be that two billion species have made their appearance on the earth since the beginning of the Cambrian period 600 million years ago."[73]

An immense number of species have lived in the past and the evidence for their extinction is graphic. An immense number of species are alive in the present; yet, Jeffrey Schwartz, Professor of Anthropology at the University of Pittsburgh, has had to admit that,

"...it was and still is the case that, with the exception of Dobzhansky's claim about a new species of fruit fly, the formation of a new species, by any mechanism, has never been observed."[74]

In summarizing the historic scenario of natural selection Tom Bethell writes in *Harper Magazine*,

"I have not been surprised to read, therefore, in Lewontin's recent book, *The Genetic Basis of Evolutionary Change* (1974), that in some of the latest evolutionary theories 'natural selection plays no role at all.' Darwin, I suggest, is in the process of being discarded, but perhaps in deference to the venerable old gentleman, resting comfortably in Westminster Abbey next to Sir Isaac Newton, it is being done as discreetly and gently as possible, with a minimum of publicity."[75]

From the very beginning of the Creation/Evolution controversy, there have been competent scientists who have found the idea of evolution by means of natural selection to be contrary to the available evidence. At the turn of the century, Vernon Lyman Kellogg of Stanford University wrote,

"The fair truth is that the Darwinian selection theory considered with regard their plain capacity to be an independently sufficient mechanical explanation of descent, stands today seriously discredited in the biological world." He went on to say no other hypothesis had replaced it. In other words, having rejected the Genesis account out of hand, Kellogg found Darwin's explanation of no value and, after scanning the horizon for other ideas, found none forthcoming that were convincing. He further said, "Mutations seem to be too few and far between. For orthogenesis we can discover no satisfactory mechanism, and the same is true for the

Lamarckian theory of modification accumulation through inheritance of acquired ontogenic character."[76]

Kellogg also noticed that one group of scientists denied in total any effectiveness or capacity for species forming on the part of natural selection while,

"While the other group...see in natural selection an evolutionary factor capable of initiating nothing, dependent wholly for any effectiveness on some primary factor or factors controlling the origin and direction of variation, but capable of extinguishable on all unadapted , unfit lines of development..."[77]

Henry Fairfield Osborn, a leading American evolutionist, also stated,

"In contrast to the unity of opinion on the *law* of evolution is a wide diversity of opinion on the *causes* of evolution. In fact the causes of evolution of life are as mysterious as the law of evolution is certain."[78]

Isn't this amazing that they are sure that the law of evolution is certain, but have no confidence in the theorized means by which it proceeds? Osborn continues,

"It may be said that Darwin's law of selection as a natural explanation of the origin of *all* fitness in form and function has also lost its prestige at the present time, and all of Darwinism which now meets with universal acceptance is the *law of the survival of the fittest,* a limited application of Darwin's great idea as expressed by Herbert Spencer."[79]

During lectures delivered at Princeton University, L.T. More asked,

"If natural selection is a force that can destroy but cannot create species and if the reasons for this destruction are unknown, of what value is the theory to mankind? ... The collapse of the theory of natural selection leaves the philosophy of mechanistic materialism in a sorry plight."[80]

More and more scientists are recognizing that natural selection may preserve, but it cannot create. Soren Lovtrup, Swedish biologist and embryologist, is one of the prominent scientists who understands this,

> "Darwin complained his critics did not understand him, but he did not seem to realize that almost everybody, friends, supporters and critics, agreed on one point, his natural selection cannot account for the origin of the variations, only for their possible survival. And the reasons for rejecting Darwin's proposal were many, but first of all that many innovations cannot possibly come into existence through accumulation of many small steps, and even if they can, natural selection cannot accomplish it, because incipient and intermediate stages are not advantageous."[81]

Michael J. Behe, Professor of Biochemistry at Lehigh University, gives us the historic perspective on the theorized value of natural selection to the theory of evolution,

> "It was a shock to the people of the 19th century when they discovered, from observations science had made, that many features of the biological world could be ascribed to the elegant principle of natural selection. It is a shock to us in the twentieth century to discover, from observations science has made, that the fundamental mechanisms of life cannot be ascribed to natural selection, and therefore were designed. But we must deal with our shock as best we can and go on. The theory of undirected evolution is already dead, but the work of science continues."[82]

In Charles Darwin's attempt to eliminate God as the source of intelligent design in nature, he, as Grasse acknowledges, simply replaced God with chance and natural selection,

> "Directed by all powerful selection, chance becomes a sort of providence, which, under the cover of atheism, is not named but which is secretly worshipped."[83]

Dr. Greg Graffin of UCLA essentially reiterates what Grasse has said above,

> "The trick is: How do you talk about natural selection without implying the rigidity of law? We use it as almost an active participant, almost like a god. In fact, you could substitute the word "god" for "natural selection" in a lot of evolutionary writings, and you'd think you were listening to a theologian. It's a routine we know doesn't exist, but we teach it anyway: genetic mutation and some active force choose the most favorable one."[84]

Darwin and his followers' attempt to essentially deify natural selection has, after being subjected to scrutiny, been found wanting just like Zeus, Isis, Apollo, and the other false gods of history. Thomas Hunt Morgan, an American evolutionary biologist, geneticist, and embryologist, is among those who acknowledge this failure,

> "Selection, then, has not produced anything new, but only more of certain kinds of individuals. Evolution, however, means producing more new things, not more of what already exists."[85]

Chapter Four End Notes

[1] George B. Johnson & Peter H. Raven, *Biology, Principles and Explorations*, Holt, Rinehart & Winston, Inc. 1996, p. 260

[2] ___, *Science Explorer, Focus on Earth Science*, Prentice Hall, 2001, p. 594

[3] K.R. Miller and J. Levine, *Biology,* Prentice Hall, 2002, p. 381

[4] Eldon Enger, Frederick Ross, & David Bailey, Concepts in Biology, 13th edition, McGraw-Hill, 2009, p. 271

[5] George B. Johnson & Peter H. Raven, *Biology*, Holt, Rinehart and Winston *Biology*, Holt, Rinehart and Winston, 2004, p.291

[6] Alton Biggs, Chris Kapicka, Linda Lundgren, *Biology, The Dynamics of Life*, Glencoe,McMillian/ McGraw-Hill, 1995, p. 432

[7] Bernice Essenfeld, Carol Gontang, Randy Moore, *Biology,* Addison-Wesley, 1994, p. 250

[8] George B. Johnson, *Biology Visualizing Life,* Holt, Rinehart, & Winston, 1994, p. 188

[9] *International Wildlife Encyclopedia*, Burton & Burton, p. 2706

[10] *Biology, The Web of Life*, Second Edition, Scott-Foresman-Addison Wesley, 2000, p.250

[11] *Ibid.*

[12] *Ibid.*

[13] *Ibid.*

[14] Kenneth R. Miller and Joseph Levine, *Biology*, Prentice Hall, 1998, p. 298

[15] George B. Johnson, *Biology Visualizing Life*, Holt, Rinehart and Winston, 1994, p. 188

[16] L. Harrison Matthews, "Introduction to Darwin's *Origin of Species*," London; J.M. Dent & Sons, Ltd., 1971 p. xi

[17] Kenneth R. Miller, Joseph Levine, *Biology*, Prentice Hall, 1991, p. 297

[18] Jerry A. Coyne, "Not Black and White," *Nature*, Vol. 36, November 5, 1998, p. 35

[19] *Ibid.*

[20] *Ibid.*

[21] *Ibid.*

[22] *Ibid.* p. 36

[23] *Ibid.*

[24] Michael E.N. Majerus, *Melanism: Evolution in Action*, Oxford University Press, 1998, p. 116

[25] *Ibid.* p. 121

[26] *Ibid.*

[27] *Ibid.* p. 121

[28] Kauri Mikkola, 1984 *Biological Journal of the Linnean Society* 21, pp. 409-421

[29] Cyril A. Clarke, *Biological Journal of the Linnean Society* 26, 1985, pp. 189-199

[30] M.W. Ho & P.T Saunders, "Beyond Neo-Darwinism--An Epigenetic Approach to Evolution," *Journal of Theoretical Biology 78,* 1979, p. 589

[31] Stephen Jay Gould, 'The Return of Hopeful Monsters,' *Natural History*, Vol. 86, June-July 1977, p. 28

[32] Niles Eldredge, *An Extravagance of Species, Natural History*, Vol. 89, No. 7, July 1980, p. 46

[33] K.R. Miller and J. Levine, *Biology,* Prentice Hall, 2002, p. 382

[34] Steven Stanley, *Proceedings of the National Academy of Science*, 72: p. 646, 1975

[35] *Ibid.*

[36] Richard Lewontin, 'Adaptation,' *Scientific American,* Vol. 239(3), September 1978, p. 159

[37] Gregory Alan Pesely, "The Epistemological Status of Natural Selection," *Laval Theologique et Philosophique*, Vol. XXXVIII, Feb. 1982, p. 74

[38] Arthur Koestler, *Janus: A Summing Up.* Vintage Books, 1978, p. 170

[39] C. Loring Brace review of *Species, Species Concepts, and Primate Evolution*, edited by William H. Kimbel and Lawrence B. Martin (Plenum Press, 1993, 560 pp.), *American Scientist*, Vol. 82, September/October 1994, pp. 484-486)
[40] Alton Biggs, Chris Kapicka, Linda Lundgren, *Biology, The Dynamics of Life*, Glencoe,McMillian/ McGraw-Hill,1995, p. 430
[41] Alton Biggs et al., *Biology: The Dynamics of Life*, Glencoe/McGraw-Hill, 2000, p. 382
[42] Alton Biggs, et al., *Biology: The Dynamics of Life*, Glencoe, 2004, p. 399
[43] Alton Biggs et al., *Biology: The Dynamics of Life*, Glencoe Science, *National Geographic*, McGraw-Hill/Glencoe, 2006, p. 399
[44] Neil A. Campbell, Brad Williamson, & Robin J. Heyden, *Biology: Exploring Life*, Pearson, 2009, p. 308
[45] Fransisco Ayala, *Scientific American*, Vol. 239, Sept. 1978, p. 63
[46] *Ibid.*
[47] *Reuters*, September 15, 1999, Retrieved from iol scitech
[48] Neil A. Campbell, Jane B. Reece, & Eric J. Simon, *Essential Biology with Physiology*, 2004, p. 245
[49] *Ibid.*, p. 241
[50] *Medical Tribune*, 12/29/88, p. 23
[51] *Ibid.*
[52] Anthony Hart, *Nature*, 9/16/99, p. 23
[53] Bernice Essenfeld, Carol Gontang, Randy Moore, *Biology*, Addison-Wesley, 1994, p. 250
[54] Roger Lewin, "Evolutionary Theory Under Fire," *Science*, Vol. 210 [4472], Nov. 21, 1980, p. 883
[55] *Modern Biology*, Holt Rinehart and Winston, 2002 p. 288
[56] Richard C. Lewontin, "Adaptation," *Scientific American*, September 1978, p. 215
[57] *Ibid.*, p. 222
[58] *Ibid.*, p. 215
[59] Richard Lewontin, "Adaptation," *Scientific American*, Sept. 1978
[60] *Ibid.*
[61] Steven Stanley, *Proceedings of the National Academy of Science*, 72: p. 646, 1975
[62] S. M. Stanley, *Pro. N. A. S.*, v 72, p. 648
[63] Tom Bethell, "Darwin's Mistake," *Harper's Magazine*, 252, Feb. 1976, pp. 70-75
[64] P.P. Grasse, *Evolution of Living Organisms*, Academic Press, 1977, p. 170
[65] S.J. Gould, *Paleobiology* 6(1):121 Winter 1980
[66] Edward Wiley and Daniel Brooks, Roger Lewin, *Science*, 1982, p. 1239-1240 (Roger Lewin, "A Downward Slope to Greater Diversity," *Science*, vol. 217 Sept 24, 1982, p. 1239
[67] Richard C. Lewontin, "Adaptation," *Scientific American*, vol. 78, p. 222
[68] Richard B. Goldschmidt, "Evolution, as Viewed by One Geneticist," *American Scientist* 40, January 1952, p. 94
[69] D. Rosen, "Darwin's Demon," in *Systematic Zoology 27*, 1978, p. 372
[70] Colin Patterson, "Cladistics," Interview on BBC, March 4, 1982
[71] Keith S. Thompson, "The Meaning of Evolution," *American Scientist*, vol. 70, September/October 1982, p. 529
[72] S.R. Palumbi, "Marine Speciation," *Annual Review of Ecology and Systematics*, 1994, p. 548.
[73] Richard Lewontin, "Adaptation," *Scientific American*, vol. 78, p. 221
[74] Jeffrey Schwartz, *Sudden Origins*, John Wiley, 1999, pg. 300
[75] Tom Bethell, "Darwin's Mistake," Harper's magazine, 1976
[76] Vernon Lyman Kellogg, *Darwinism to-day*, Henry Holt and Company, 1908, p. 5
[77] *Ibid*, p. 27
[78] Henry Fairfield Osborn, *The Origin and Evolution of Life*, Charles Scribner's Sons, 1918, p. ix
[79] *Ibid*, p. xv
[80] L.T. More, *The Dogma of Evolution*, Princeton University Press, 1925, p. 240

[81] Soren Lovtrup, *Darwinism: The Refutation of a Myth*, Croom Helm, 1987, pp. 274-275
[82] Michael J. Behe, "Molecular Machines," *Cosmic Pursuit*, Spring 1998, p. 35
[83] P.P. Grasse, *Evolution of Living Organisms*, 1977, p. 107
[84] Greg Graffin (in an interview with David Biello), "Was Darwin a Punk? A Q&A with Punker-Paleontologist Greg Graffin," *Scientific American*. 303 (5): 28, 2010
[85] Thomas Hunt Morgan & Louis Clark Vanuxem Foundation, *A Critique of the Theory of Evolution*, Princeton University Press, 1919, p. 154

VESTIGIAL ORGANS

Introduction

The term, "vestigial organs," is used by the evolutionary establishment in an effort to convince students that the human body possesses a variety of organs and structures which have no purpose. Presumably, these organs and structures were once very useful in one of our various evolutionary animal ancestors. Approximately a hundred years ago, about 180 of these organs and structures were listed as vestigial. At the present, this long list has shrunken to almost zero. If there had been more humility and less arrogance, the scientists would have taken the position that it is more than likely that the Creator put these organs in the human body for a purpose. Therefore, ongoing research will most likely discover the value of these organs. However, since God was written out of the picture, these men arbitrarily and prematurely relegated the organs and structures whose purpose, at the time they did not understand, to the category of vestigial organs. Among these, were such important structures as the pineal gland, pituitary gland, thymus gland, tonsils, appendix, and coccyx (tailbone). Today, the high-school textbooks still tenaciously cling to this outmoded idea. Generally speaking, they continue to list three or four organs as vestigial such as the appendix and the coccyx.

Textbook Evolution

> "Figure 18.10 - Vestigial structures provide evidence of evolution because they show structural change over time. The small, wormlike appendix has reduced function in humans. It is a remnant of the digestive pouch known as the caecum, which is used by plant-eating animals for digestion of cellulose."[1] (Alton Biggs, Chris Kapicka, Linda Lundgren, *Biology, The Dynamics of Life*, Glencoe, McMillian/ McGraw-Hill)

> "Many organisms have features that seem to serve no useful function. For example, humans have a tailbone at the end of the spine that is of no apparent use. The human appendix, a small, fingerlike projection from the intestine, also has no known function."[2] (*Modern Biology*, Holt, Rinehart and Winston)

"We may not have any use for our appendixes.... These structures, however, are still useful to related animals. Thus, the presence of vestigial structures is evidence of a common ancestor for us and related animals. We still have the genes for appendixes and third eyelids, even though we don't use these structures."[3] (Kaskel, Albert, Paul J. Hummer, Jr., Lucy Daniel, *Biology: An Everyday Experience*, Glencoe/McGraw-Hill)

"Further evidence of evolution is found in structures called vestigial structures that seem to serve no function but that resemble structures with functional roles in related organisms. For example, the human tailbone, or coccyx, is made up of four fused vertebrae that resemble the bones in an animal's tail. Other examples of vestigial structures are the pelvic bones of modern whales and the human appendix."[4] (John H. Postlethwait & Janet L. Hopson, *Modern Biology*, Holt, Rinehart, & Winston)

"A structure becomes vestigial when the species no longer needs the feature for its original function, yet it is still inherited as part of the body plan for the species."[5] (Alton Biggs, et. al., *Biology: The Dynamics of Life*, Glencoe Science, *National Geographic*, McGraw-Hill/Glencoe)

Science Versus

Contrary to evolutionary propaganda, essentially all of the so-called vestigial organs have been found to be quite useful in modern human beings. One of the last few hold-outs in support of the evolutionary argument has been the appendix. Writers of high school textbooks are taken up with the idea that it is essentially valueless. However, research has shown that it plays an important part in the human immune system. A number of writers have contributed to our understanding of this function of the appendix. From Frederic Martini in his book, *Fundamentals of Anatomy & Physiology*, we learn that the appendix is part of the lymphatic system,

"The mucosa and submucosa of the appendix are dominated by lymphoid nodules, and the appendix's primary function is as an organ of the lymphatic system."[6]

As a reminder to the reader of the function and value of the lymphatic system, I would like to reference Kim Ann Zimmermann who informs us that,

> "The primary function of the lymphatic system is to transport lymph, a clear, colorless fluid containing white blood cells that helps rid the body of toxins, waste and other unwanted materials."[7]

Kenneth Saladin of Georgia College and State University reports,

> "The appendix is densely populated with lymphocytes [a type of white blood cell] and is a significant source of immune cells."[8]

From Van de Graff and Fox in their book, *Concepts of Human Anatomy & Physiology,*

> "The appendix contains masses of lymphoid tissue that may serve to resist infection."[9]

Roy Hartenstein writing in the *Grolier Encyclopedia* states,

> "Long regarded as a vestigial organ with no function in the human body, the appendix is now thought to be one of the sites where immune responses are initiated."[10]

Finally, Matt Cartmill of Duke University along with Hylander and Shafland comment,

> "The human appendix has masses of lymphatic tissue in its walls and seems to provide a local defense against infection from micro-organisms in the colon."[11]

> "People sometimes speak of it as a vestigial organ, as though it were a useless remnant of a long caecum like that of a rabbit. It is not."[12]

With regard to "vestigial organs" in general, evolutionary biologist, S.R. Scadding, admits,

> "An analysis of the difficulties in unambiguously identifying functionless structures...leads to the conclusion that 'vestigial organs' provide no evidence for evolutionary theory."[13]

Now, with regard to the evolutionary idea that the hip bones in whales are vestigial organs, which are left over from a theoretical land-dwelling ancestor, let's take a closer look. First, a review of the textbook, *Biology: The Dynamics of Life*,

> "Vestigial structures, such as pelvic bones in the baleen whale, are evidence of evolution because they show structural change over time."[14]

Above is what the high school textbooks state. Now, by contrast, let's review the work of Everhard Johannes Slijper who was professor of general zoology at Amsterdam University, Netherlands. He has long been regarded as the world's expert on whales. Concerning these bones, which he referred to as 'pelvic bones,' he stated, "...unlike the pelvis of normal mammals it is not attached to the vertebral column."[15] Actually, these bones are necessary to facilitate reproduction.

Chapter Five End Notes

[1] Alton Biggs, Chris Kapicka, Linda Lundgren, *Biology, The Dynamics of Life*, Glencoe, McMillian/ McGraw-Hill, 1995, p. 433
[2] *Modern Biology*, Holt, Rinehart and Winston, 2002, p. 290
[3] Kaskel, Albert, Paul J. Hummer, Jr., Lucy Daniel, *Biology: An Everyday Experience*, Glencoe McGraw-Hill, 2003, pg. 621
[4] John H. Postlethwait & Janet L. Hopson, *Modern Biology*, Holt, Rinehart, & Winston, 2006, p. 305
[5] Alton Biggs, et. al., *Biology: The Dynamics of Life*, Glencoe Science, *National Geographic*, McGraw-Hill/Glencoe, 2006, p. 401
[6] Frederic Martini, *Fundamentals of Anatomy & Physiology*, Prentice Hall, 1998, p. 899
[7] Kim Ann Zimmermann, *Lymphatic System: Facts, Functions & Diseases*, February 08, 2013, LiveScience website
[8] Kenneth Saladin, *Anatomy & Physiology*, McGraw Hill, 2001, p. 974
[9] Kent Marshall Van de Graff & Stuart Ira Fox, *Concepts of Human Anatomy & Physiology*, 1999, p. 837
[10] Roy Hartenstein, *Grolier Encyclopedia*, 2002, Grolier Interactive, Inc.
[11] Matt Cartmill, William L. Hylander and James Shafland, *Human Structure*, 1987, p. 135
[12] Matt Cartmill, William L. Hylander and James Shafland, *Human Structure*, 1987, p. 136
[13] S.R. Scadding, "Do Vestigial Organs Provide Evidence for Evolution?" Evolutionary Theory, vol. 5 may 1981, p. 173
[14] Alton Biggs, et al., *Biology: The Dynamics of Life*, Glencoe Science, *National Geographic*, McGraw-Hill/Glencoe, 2006, p. 402
[15] E.J. Slijper, *Whales*, translated from Dutch by A. J. Pomerans, Cornell University Press, 2nd ed., 1979

EMBRYOLOGY

Introduction

The opposition to the concept that the human embryo in it's development, passes through the general stages of mankind's theoretical, evolutionary history has been prominent in scientific literature from the very inception of this idea to the present. The catch phrase invented by the German embryologist, Ernst Haeckel, "ontogeny recapitulates phylogeny," does not reflect the true nature of human embryological development. It is a statement that is untrue, has been proven to be untrue repeatedly by a variety of competent embryologists, and has been repeatedly reported in scientific literature. Nevertheless, because Darwin's loyal followers have found this concept to be attractive as a support for the theory of evolution, high school biology textbooks continue to present it in one way or another as if somehow it still retains some measure of validity. First, let us briefly review a variety of biology texts which continue to present this concept as a support for the theory of evolution.

Textbook Evolution

> "Figure 15.10. Comparing embryos can reveal their evolutionary relationships. The presence of gill slits and tails in early vertebrate embryos shows that they may share a common ancestor."[1] (*Biology: The Dynamics of Life*, Glencoe/McGraw-Hill, 2000, p. 410)

> "The early stages of different vertebrate embryos are strikingly similar to each other... The German zoologist Ernst Haeckel (1934-1919), who was also struck by these similarities, declared that "ontogeny recapitulates phylogeny." This statement can be translated to "embryological development repeats evolutionary history."[2] (*Modern Biology*, Holt, Rinehart and Winston, 2002, p. 290)

> "Nevertheless, the similarities in early embryonic stages of vertebrates can be taken as yet another indication that vertebrates share a common ancestry."[3] (*Ibid.*, p. 291)

"There have, in the past, been incorrect explanations for these similarities. Also, the great biologist Ernst Haeckel fudged some of his drawings to make the earliest stages of some embryos seem more similar than they actually are!"[4] (*Ibid.*)

"Errors aside, however, it is clear that the same groups of embryonic cells develop in the same order and in similar patterns to produce the tissues and organs of all vertebrates. These common cells and tissues, growing in similar ways, produce the homologous structures discussed earlier."[5] (*Ibid.*)

"Another comparison also provides evidence of evolution. Figure 29-13 shows five different animal embryos. Fish, frog, turtle, bird, and rabbit embryos are shown. Which embryo is which? You really can't tell. They all look very much alike at this stage in their development. As they get older, you will be able to tell which embryo is which. How is the similar appearance of embryos evidence of evolution? All five animals are chordates. They have a common ancestry. The embryos share some of the same traits from their common ancestry. These traits result in all the embryos looking similar."[6] (*Biology: An Everyday Experience*, Glencoe/McGraw-Hill, 2003, p. 620)

"At this stage of development, all the embryos have a tail and pharyngeal pouches. In fish, these pouches develop into the supports for the gills, while in mammals, reptiles, and birds, they develop into parts of ears, jaws, and throat. It is the shared features in the young embryos that suggest evolution from a distant, common ancestor."[7] (*Biology: The Dynamics of Life*, Glencoe, 2004, p. 402)

"Scientists make inferences about evolutionary relationships by comparing how different species develop before birth. During early development, all vertebrates have a tail and rows of tiny slits in their throats. These similarities suggest that vertebrate species share a common ancestor."[8] (*California Focus on Life Science*, Pearson/Prentice Hall, 2008, p. 235)

"Vertebrate include fishes, amphibians, reptiles, birds, and mammals. Yet despite how different these groups are, comparing the ways in which their embryos develop provides strong evidence of their evolutionary links."[9] (Cecie Starr & Beverly McMillan, *Human Biology*, ninth edition, Brooks/Cole, Cengage Learning, 2010, p. 450) (College Textbook)

Science Versus

The importance of Haeckel's contribution to the theory of evolution has been well stated by Stephen Gould of Harvard,

> "Haeckel was the chief apostle of evolution in Germany. Nordenskiold (1929) argues that he was even more influential than Darwin in convincing the world of the truth of evolution. Yet influential as Haeckel was among scientists, his general impact was even greater...His major popular work *Weltratsel* ("The Riddle of the Universe"; 1899), was among the most spectacular successes in the history of printing. It sold 100,000 copies in tis first year, went through ten editions by 1919, was translated into twenty-five languages, and had sold almost half a million copies in Germany alone by 1933."[10]

The many quotes listed below will make it clear that the human embryo does not pass through a series of anatomical and physiological stages which reflect a long, evolutionary pathway from fish to humans. It has, for instance, been fully demonstrated that the human embryo does not have gill slits. In the neck area of the fish embryo there are barbs and grooves, some of which open into the throat. These, in fact, do become slits, which soon develop into gills. However, in the human embryo the neck region contains little grooves, which are technically referred to as pharyngeal pouches. None of these open into the throat and none become slits, let alone gills. They do develop into the lower jaw, and certain glands, such as the thymus gland.

Further evidence against the theory of recapitulation is to be seen in the observable facts that, while in lower animals, teeth appear before tongues, in human's, tongues appear before teeth as the embryo develops.

The idea that the kidney development also represents the phylogenetic pattern is debunked by Dr. Alfred S. Romer, American paleontologist and specialist in vertebrate evolution. He states,

> "We see in this story the development in the amniote embryo a three, successive kidney structure pronephros, mesonephros and metanephros. It is often stated or implied that these three are

> distinct kidneys which have succeeded one another phylogenetically as they do embryologically. Upon consideration, however, it will be seen that there is no strong reason to believe this. The differences are readily explainable on functional grounds."[11]

One of the major criticisms against this theory is that other researchers have found that Haeckel actually misrepresented the development of the human embryo. This has been reported regularly in the science literature for a long time. For instance, W. R. Thompson, Director of the Commonwealth Institute of Biological Control at Ottawa, Canada, writes,

> "When the 'convergence' of embryos was not entirely satisfactory, Haeckel altered the illustrations of them to fit his theory. The alterations were slight but significant. The 'biogenetic law' as a proof of evolution is valueless."[12]

Although it was previously referenced in a high school textbook above, it is important to use yet another high school text (*Biology*, Prentice Hall) in order to show how far evolutionists will go to protect a bad theory,

> "There have, in the past, been incorrect explanations for these similarities. Also, the biologist Ernst Haeckel fudged some of his drawings to make the earliest stage of some embryos seem more similar than they actually are!"[13]

To say Haeckel simply "fudged" on his drawings is a preposterous statement. The man was grossly dishonest and his errors were purposive. This was not a light-hearted matter, as Gavin deBeer acknowledges,

> "...the theory of recapitulation has had a great and, while it lasted, regrettable influence on the progress of embryology."[14]

In 1987, historian Jane Oppenheimer reminded us that Haeckel's,

> "...hand as an artist altered what he saw with what should have been the eye of a more accurate beholder. He was more than once, often justifiably, accused of scientific falsification, by William His and many others."[15]

Michael K. Richardson is a researcher and lecturer at the Institute of Biology Leiden, The Netherlands. In his very detailed study he reports that Haeckel's drawings were,

> "...not consistent with other data on the development of these species." He says that "...these famous images are inaccurate and give a misleading view of embryonic development."[16]

Again, Michael K. Richardson along with his colleagues, reveal,

> "Our survey seriously undermines the credibility of Haeckel's drawings."[17]

Harvard's Stephen Gould also faults Haeckel, stating that he,

> "...exaggerated the similarities by idealizations and omissions," Gould further speaks of them as, "inaccuracies and outright falsification."[18]

In an interview with Elizabeth Pennisi, Richardson is quoted as saying,

> "It looks like it's turning out to be one of the most famous fakes in biology."[19]

Now let's take a closer look at one of the biology text statements (*Biology: The Dynamics of Life*, Glencoe/McGraw-Hill),

> "You know that reptiles, birds, and mammals do not have gills when they are mature. As development continues, the differences in the embryos will increase until you can distinguish among them. However, the similarities among the young embryos suggest evolution from a distant, common ancestor."[20]

Note the word, "until." This word implies an indefinite (or even extensive) length of time in which the embryos are very similar in appearance. However, from the very beginning competent scientists such as Adam Sedgwick, one of the founders of modern geology, writing in 1894, strongly opposed the implied idea that human and a variety of animal embryos were difficult to distinguish from one another during their earliest period of development,

"There is no stage of development in which the unaided eye would fail to distinguish between them with ease....not necessary to emphasize further these embryonic differences."[21]

Sedgwick is emphatic that there is no significant time period in which these embryos, referenced by Haeckel, remain indistinguishable from one another,

"...I can distinguish a fowl and a duck embryo on the second day....every embryologist knows that they exist and could bring forward innumerable instances of them. I need only say with regard to them that *a species is distinct and distinguishable from its allies from the very earliest stages all through development*."[22] (emphasis in original)

Years after Segdewick made his comments attacking Haeckel's work, Erich Blechschmidt, the German embryologist and anatomist, also argued that,

"The early stages of human embryonic development are distinct from the early development of other species."[23]

In 1987, Richard Elinson, a professor within the Department of Biological Sciences at Duquesne University, stated that frogs, chicks, and mice,

"...are radically different in such fundamental properties as egg size, fertilization mechanisms, cleavage patterns, and [gastrulation] movements."[24]

Three of the McGraw-Hill texts, two by Peter Raven and George B. Johnson and the third by Sylvia S. Mader, also emphasize the similarities in the embryos during the "early" stages of development,

"Notice that the early embryonic stages of these vertebrates bear a striking resemblance to each other."[25]

"Some of the strongest anatomical evidence supporting evolution comes from comparison of how organisms develop. In many cases, the evolutionary history of an organism can be seen to unfold during its

> development, with the embryo exhibiting characteristics of the embryos of its ancestors."[26]

> "At these comparable early developmental stages, the two have many features in common, although eventually they are completely different animals. This is evidence that they evolved from a common ancestor."[27]

This is nonsense. As Sedgwick and others have shown, embryos are known to be quite different at their early stages. It is actually at a midway through development point or stage that they do show some similarities. Then, as they progress through the midway into the latter stage of development, they become more and more dissimilar. Haeckel in his deceptive drawings, used the midway point for the first stage of his drawings.

As stated before, numbers of authorities in this area have clarified that the human embryo does not have gill slits or gills. For instance, Louis Wolpert, a developmental biologist, clarifies,

> "A higher animal, like the mammal, passes through an embryonic stage when there are structures that resemble the gill clefts of fish. But this resemblance is illusory and the structures in mammalian embryos only resemble the structures in the *embryonic* fish that will give rise to gills."[28]

This is also affirmed by Dr. Gunter Rager of the Institute of Anatomy and Special Embryology at University Fribourg,

> "The concept 'pharyngeal arches' is purely descriptive and ideologically neutral. It describes folds which appear [in the neck] region....In man, however, gills do never exist."[29]

Over time, the weight of the evidence has made a small dent in the writing of biological textbooks. In some cases the textbook writers have backed away from blatant statements as were more prominent in the textbooks of the 1960s, 70s and 80s. Nevertheless, they are quite unwilling to let the whole concept die away. They are simply restating it. They have abandoned the idea that the human embryo recapitulates the adult stages of its evolutionary ancestors. Instead, they are now suggesting that the human embryo recapitulates the embryological stages. Since these embryos are so small at the beginning it is to be expected that there would be some similarities in "appearance." After all, they all start out as a single cell. However, an embryologist who, today, cannot tell the difference between a fertilized human egg and a fertilized fish egg is totally incompetent.

Further, if human embryos really did recapitulate those of fish, frogs, reptiles and mammals, then we should see something quite different in development of the embryos that we do see. Sir Gavin de Beer, one of the top evolutionary spokesmen of the past, shows in his book, *Homology, An Unsolved Problem*, the improbability of such ideas. He points out that in the nuke (amphibian) the forelimbs develop from trunk segments 2 to 5, in the lizard (a reptile) 6 to 9, but in humans the forelimbs develop from trunk sections 13 to 18. What is there in this that shows embryological recapitulation? Absolutely nothing.

So, the textbook writers seem to be determined, one way or another, to keep this fraudulent idea that used to be so valuable to the theory of evolution, in the textbooks as a means of convincing the young students that we are the product of a long, naturalistic, evolutionary history.

Although science textbooks may not be showing much respect for Richardson's research in this area, various science periodicals are giving it serious attention. Elizabeth Pennisi reports,

> "Not only did Haeckel add or omit features, Richardson and his colleagues report, but he also fudged the scale to exaggerate similarities among species, even when there were ten-fold differences in size. Haeckel further blurred differences by neglecting to name the species in most cases, as if one representative was accurate for an entire group of animals."[30]

Prior to our concluding remarks in this category, it will be of interest to quote from a variety of places in Richardson's article which represents the most recent and comprehensive research on this subject that has been made to date. Haeckel, in his deceptive drawings, portrayed the vertebrate embryos as being the same size. By contrast, Richardson and his colleagues found that they differed significantly,

> "Contrary to recent claims that all vertebrate embryos pass through a stage when they are the same size, we find a greater than 10-fold variation in greatest length at the tailbud stage. Our survey seriously

undermines the credibility of Haeckel's drawings, which depict not a conserved stage for vertebrates, but a stylised amniote embryo."[31]

The words "stylised amniote embryo" are simply a nice way of saying that Haeckel customized his drawings in order to give support to the theory of evolution.

This deceptive practice was also noted by the late professor, Richard Goldshmidt, a geneticist previously with the University of California at Berkeley,

> "The present generation cannot imagine the role he played in his time, far beyond his actual scientific performance...Haeckel's easy hand at drawing made him improve on nature and put more into the illustrations than he saw...one had the impression that he first made a sketch from nature and then drew an ideal picture as he saw it in his mind."[32]

Haeckel's drawings were intended to demonstrate his recapitulation theory of biogenetic law. Michael Richardson et al. explains,

> "A conserved stage was a necessary part of his theory because evolution was claimed to progress principally by the terminal addition of new adult stages to the end of ancestral developmental sequences."[33]

> "Haeckel's ideas soon came in for strong criticism. His drawings are also highly inaccurate, exaggerating the similarities among embryos, while failing to show the differences."[34]

> "...and Richardson (1995) have argued that Haeckel's drawings are inaccurate, and we have now provided persuasive evidence that this is indeed the case."[35]

Summary

In conclusion, we have some straightforward remarks listed below which should, if objective rather than subjective science was being employed in this argument, put the matter to rest once and for all. This foolish theory has been opposed, as shown above, by Sedgwick (Darwin's mentor) and others from the very beginning. Note anthropologist Ashley Montagu's reference to this idea being falsified as far back as 1921,

> "The theory of recapitulation was destroyed in 1921 by Professor Walter Garstang in a famous paper. Since then no respectable biologist has ever used the theory of recapitulation, because it was utterly unsound, created by a Nazi-like preacher named Haeckel."[36]

Further, Conrad Hal Waddington, FRS, a paleontologist and biologist, states,

> "The type of analogical thinking which leads to theories that development is based on the recapitulation of ancestral stages or the like no longer seems at all convincing or even interesting to biologists."[37]

George G. Simpson, one of the most influential paleontologists of the 20th century and his associate, W. Beck, are emphatic that,

> "Haeckel misstated the evolutionary principle involved. It is now firmly established that ontogeny does not repeat phylogeny."[38]

Stephen Gould affirms,

> "...the theory of recapitulation should be defunct today"[39]

R.H. Dott and R.L. Batten, in their college text, also refute Haeckel's idea,

> "Much research has been done in embryology since Haeckel's day, and we now know there are too many exceptions to this analogy, and that ontogeny does not reflect accurately the course of evolution."[40]

Keith S. Thompson, in his *American Scientist* article, "Ontogeny and Phylogeny Recapitulated," says,

> "Surely the biogenetic law is as dead as a doornail. It was finally exorcised from biology textbooks in the fifties. As a topic of serious theoretical inquiry, it was extinct in the twenties."[41]

The Biogenetic law has been dear to the hearts of textbook writers for a very long time. Because it appears (if it were true) to be such a good support for the theory of evolution, they are quite reluctant to let it go. Dr. Walter J. Bock of the Department of Biological Sciences at Columbia University comments on this,

> "Moreover, the Biogenetic law has become so deeply rooted in biological thought that it cannot be weeded out in spite of its having been demonstrated to be wrong by numerous scholars."[42]

The propaganda points of evolution, when subjected to scrutiny, sooner or later fall apart. This is certainly the case with regard to the recapitulation theory, as Biologist and educator, Paul R. Ehrlich, and his associate, Richard W. Holm, illustrate,

> "This generalization was originally called the biogenetic law by Haeckel and is often stated as 'ontogeny [the development of an embryo] recapitulates [repeats] phylogeny [evolution].' This crude interpretation of embryological sequences will not stand close examination, however. Its shortcomings have been almost universally pointed out by modern authors, but the idea still has a prominent place in biological mythology."[43]

In their book, *Concepts of Zoology*, Hubert and Marie Frings, write,

> "...we no longer believe we can simply read in the embryonic development of a species its exact evolutionary history."[44]

Further, Richardson et. al reports that,

> "One puzzling feature of the debate in this field is that while many authors have written of a conserved embryonic stage, no one has cited any comparative data in support of this idea. It is almost as though the phylotypic stage is regarded as a biological concept for which no proof is needed."[45]

In other words, this idea that the human embryo repeats its long evolutionary history has long ago passed out of the realm of being a mere scientific theory. It has now become a politically correct theory for which no further proof is needed. This kind of thinking applies to the whole theory of evolution. Why? Because the theory of evolution provides for the current culture a necessary support for a preferred worldview. Any evidence to the contrary is relegated to the inconsequential, and any flaws or fraudulent aspects in the theory are excused on the basis that the theory must be correct and that "apparent" flaws will be resolved in due time.

Nevertheless, with regard to the so-called biogenetic law, Richardson, in an interview with Nigel Hawks of *The Times* in London, referred to Haeckel as an "embryonic liar." Hawks further stated that,

> "This is one of the worst cases of scientific fraud. It's shocking to find that somebody one thought was a great scientist was deliberately misleading. It makes me angry... What he [Haeckel] did was to take a human embryo and copy it, pretending that the salamander and the pig and all the others looked the same at the same stage of development. They don't... These are fakes."[46]

With all of this overwhelming evidence against the idea that similarity amongst various creatures during their embryological development is a strong support for evolution, the textbooks, unwilling to let go of this idea, still continue to use it by simply adding a modern modification of the old idea,

> "Figure 15-17 In their early stages of development, chickens, turtles, and rats look similar, providing evidence that they shared a common ancestry."[47] (*Biology*, Prentice Hall, 2004)

From the Piltdown Man to the fraudulent drawings of Haeckel and from these to the fabricated peppered moth experiment of Kettlewell, you have the kind of "science" that has historically supported the theory of evolution.

Chapter Six End Notes

[1] *Biology: The Dynamics of Life*, Glencoe/McGraw-Hill, 2000, p. 410
[2] *Modern Biology*, Holt, Rinehart and Winston, 2002, p. 290
[3] *Ibid.*, p.291
[4] *Biology*, Prentice Hall, 2002, p. 385
[5] *Ibid.*
[6] *Biology: An Everyday Experience*, Glencoe/McGraw-Hill, 2003, p. 620
[7] *Biology: The Dynamics of Life*, Glencoe, 2004, p. 402
[8] *California Focus on Life Science*, Pearson/Prentice Hall, 2008, p. 235
[9] Cecie Starr & Beverly McMillan, *Human Biology*, ninth edition, Brooks/Cole, Cengage Learning, 2010, p. 450
[10] Stephen Gould, *Ontogeny and Phylogeny*, 1977, p. 77
[11] Alfred S. Romer, *The Vertebrate Body*, W.B. Saunders Company, 1965, pp. 283-284
[12] W.R. Thompson, "Introduction" to *Origin of Species*, by Charles Darwin, Everyman's Library, Dutton, 1956, p. 12
[13] Kenneth R. Miller & Joseph S. Levine, *Biology*, Prentice Hall, 2004, p. 385
[14] Gavin R. deBeer, *Embryos and Ancestors*, revised edition, Oxford University Press, 1951, p. 10
[15] Jane M. Oppenheimer, "Haeckel's Variations on Darwin," in Henry M. Hoenigswald and Linda F. Wiener [editors] *Biological Metaphor and Cladistic Classification*, University of Pennsylvania Pr, 1987, p. 134
[16] Michael K. Richardson, "Heterochrony and the Phylotypic Period," *Developmental Biology* 172, 1995, p. 412-421
[17] Michael K. Richardson, et al., "Somite Number and Vertebrate Evolution," *Development* 125, 1998, pp. 151-160
[18] Stephen J. Gould, *"Abscheulich!," Natural History,* 2000, p. 42-49
[19] Richardson in an interview with Elizabeth Pennisi, "Haeckel's Embryos: Fraud Rediscovered," *Science* Vol. 277, 5 September 1997, p. 1435
[20] *Biology: The Dynamics of Life*, Glencoe/McGraw-Hill, 2000, p. 410
[21] Adam Sedgwick, "On the Law of Development commonly known as von Baer's Law; and on the Significance of Ancestral Rudiments in Embryonic Development," *Quarterly Journal of Microscopical Science* 36, April 1, 1894, p. 35-52
[22] *Ibid.*
[23] Erich Blechschmidt, *The Beginnings of Human Life*, translated by Transemantics, 1977, p. 29-30
[24] Richard P. Elinson, "Change in developmental patterns: embryos of amphibians with large eggs," *Development as an Evolutionary Process*, vol. 8, edited by R.A. Raff and E.C. Raff, p. 3, 1987
[25] Peter H. Raven and George B. Johnson, *Biology*, Fifth edition, WCB/McGraw-Hill, 1999, p. 1181
[26] *Ibid.*, p. 416
[27] Sylvia Mader, *Biology*, Sixth edition, WCB/McGraw-Hill, 1998, p. 298
[28] Lewis Wolpert, *The Triumph of the Embryo*, Oxford University Press, 1991, p. 12
[29] Gunter Rager, "Human Embryology and the Law of Biogenesis," *Rivista di Biologia* 79, 1986, p. 449-465
[30] Elizabeth Pennisi, *Science*, vol. 277, 9/5/97
[31] Michael K. Richardson, James Hanken, Mayoni L. Goonernatne, Claude Pleau, Albert Raynaud, Lynne Selwood, Glenda M. Wright. *Anatomy and Embryology*, 1997, 196:91-106, p. 91
[32] Richard B. Goldshmidt, *The Golden Age of Zoology: portraits from memory*, University of Washington Press, 1956, pp. 31-33
[33] Michael K. Richardson et al., *There is No Highly Conserved Embryonic Stage in the Vertebrates: Implications for Current Theories of Evolution and Development*, 196 Anatomy and Embryology, p. 92, 1997
[34] *Ibid.*
[35] *Ibid.*, p. 104

[36] Ashley Montagu, quoted by Luthor D. Sunderland in *Darwin's Enigma*, Master Book Publishers, 1984, p. 119
[37] Conrad Hal Waddington, *Principles of Embryology*, George Allen and Unwin Ltd., 1956, p. 10
[38] G.G. Simpson and W. Beck, *An Introduction to Biology,* Harcourt Brace and World, 1965, p. 273
[39] Stephan J. Gould, "Dr. Down's Syndrome," *Natural History*, April 1980, p. 144
[40] R.H. Dott and R.L. Batten, *Evolution of the Earth,* 3rd Ed., 1981, p. 216
[41] Keith S. Thompson, "Ontogeny and Phylogeny Recapitulated," *American Scientist*, Vol. 76, May/ June, 1988, p. 273
[42] Walter J. Bock, "Evolution by Orderly Law," *Science*, Vol. 164, May 9, 1969, pp. 684-685
[43] Paul R. Ehrlich and Richard W. Holm, *The Process of Evolution*, McGraw-Hill, 1963, p. 66
[44] Hubert Frings and Marie Frings, *Concepts of Zoology*, Macmillan Publishing Co., 1970, p. 267
[45] Michael K. Richardson et al., *There is No Highly Conserved Embryonic Stage in the Vertebrates: Implications for Current Theories of Evolution and Development*, 196 Anatomy and Embryology, p. 92, 1997
[46] Nigel Hawkes, *The Times* (London), p. 14, August 11, 1997
[47] Kenneth R. Miller & Joseph Levine, *Biology*, Prentice Hall, 2004, p. 385

THE AGE OF THE EARTH

Textbook Evolution

The importance of long ages to the theory of evolution is stated boldly by Lawrence Badash, Professor of History of Science, at the University of California, Santa Barbara,

> "As the sun's first ray's of thermonuclear light blazed across the galaxy 4.5 billion years ago, the primal earth emerged from a spinning, turbulent cloud of gas, dust and planetoids that surrounded the new star. ...On these figures for the age of the earth rest all of geology and evolution."[1] (Lawrence Badash, *Scientific American*, Aug., 1989, p. 90)

The high school textbooks tell the students,

> "Figure 21-1 Earth's long history is divided into specific units of time in the geologic time scale."[2] (*Earth Science: Geology, the Environment, and the Universe* McGraw Hill/Glencoe)

> "Periods are defined by the life-forms that were abundant or became extinct during the time in which specific rocks were deposited."[3] (*Ibid.*, p. 556)

> "Radioactive dating is used to determine the absolute ages of rocks. In radioactive dating, scientists first measure the amount of a radioactive element (the "parent" element) in a mineral rock. Then they measure the amount of the stable element into which the radioactive element decays (the "daughter" element). By calculating the ratio of the radioactive element to the stable element, scientists can determine the age of the rock."[4] (*California Life Science,* Pearson/Prentice Hall)

Actually, radioactive dating has almost nothing to do with the dating of the rocks. As we will show in the following pages, it is merely an attempt to give the highly subjective evolutionary method of dating the rocks a boost from the area of science.

Circular Reasoning

One weak element in the business of dating the earth in support of evolution has been the phenomenon of circular reasoning: the dating of the rocks by the fossils and the dating of the fossils by the rocks. The quote below is a prime example of this concept. This is from an in-house publication (possibly by the Smithsonian), Mesozoic Vertebrate Life #1, "The Iren Dabasu Fauna from Mongolia" by Michael K. Brett-Surman of the Department of Geology at George Washington University,

> "The fauna itself was dated lower Upper Cretaceous; it lies at the base of the Upper Cretaceous time period, about 90-100 million years ago. This was done by dating the rocks according to the morphological complexity of the animals in the formation. The more "morphologically advanced" the animals were, the younger are the rocks relative to other formations with "less advanced" animals. A favorite saying of paleontologists is, 'You date the rocks by the fossils, and the fossils by the rocks.' There is no exception here."

Science Versus

The information below represents a large body of evidence, which, taken together, demonstrates that the "millions of years" age of the earth idea, while popular and while maintained by consensus, is contrary to the evidence from a variety of areas.

The Dating of Rocks

The historic development of dating was all developed long before the invention of radiometric dating techniques. Therefore, the primary means of establishing these very old dates was by an estimation value comparing sediments and fossils.

John Reader, in his book entitled, *Missing Links*, acknowledges,

> "Estimates of fossils' actual ages frequently appeared in books and articles, it is true, but they were very rough estimates based on guesses

of how fast evolution proceeds, or how long sedimentary deposits take to accumulate."[5]

First, it might be important to let the reader know how rocks are not dated.

1. Rocks are not dated by means of their structural features. The "ages" of the rocks are not necessarily identified by any physical features or breaks in the structures of the rocks. As J.A. Jeletzsky, writing in *Bulletin American Association of Petrolium Geologists,* declares,

   "It is, indeed, a well established fact that the (physical-stratigraphical) rock units and their boundaries often transgress geologic time planes in most irregular fashion even within the shortest distances."[6]

2. The adjacent relationship of rocks does not necessarily determine the ages of the rocks. Rocks of any age may lie on top of those of any other age. In his article, "Mountain-Building Chronology and the Nature of the Geologic Time Scale," Edmund M. Spieker relates,

   "Further, how many geologists have pondered the fact that lying on the crystalline basement are found from place to place not merely Cambrian, but rocks of all ages."[7]

3. The position of rocks on top of one another does not provide a reliable guide to the age of the rocks. Regarding this concept, the high school textbook by Glencoe, *Earth Science: Geology, the Environment, and the Universe,* addresses the subject as follows,

   "The principle of original horizontality states that sedimentary rocks are deposited in horizontal or nearly horizontal layers. While we may not know the actual ages of the rock layers, we can assume that the oldest rocks are at the bottom and that each successive layer going toward the top is younger."[8]

   Referring to an illustration in the book (Figure 21-5), the text goes on to say,

> "This is an application of the principle of superposition, which states that in an undisturbed rock sequence, the oldest rocks are at the bottom and each successive layer is younger than the layer beneath."[9]

Ordinarily, as stated above, the deposition of sedimentary layers of rocks would insure that the lowest strata should be the oldest and the upper layers the youngest. However, since the fossil groups often do not always fit this neat arrangement within the successive layers, this reasonable guide has not been reliable.

4. The popularity of the phenomenon of radiometric dating seems to imply to the public that radiometric dating is the sure-fire means of dating the earth. The statements below are how the Glencoe high school text discusses this topic,

> "The emission of radioactive particles and the resulting change into other elements over time is called radioactive decay. Once the emission of these atomic particles begins, the rate remains constant regardless of environment, pressure, temperature, or any other physical changes. Thus, these atomic particles become accurate indicators of the absolute age of an object."[10]

The high school text continues,

> "In a process called radiometric dating, scientists attempt to determine the ratio of parent nuclei to daughter nuclei within a given sample of a rock or fossil. This ratio is then used to determine the absolute age of the rock or fossil."[11]

Strange as it may seem, the age of rocks is not primarily determined by radiometric dating. As we have previously mentioned, the geologic column and the theorized ages were well established long before radioactive dating methods were discovered. Spieker offers a detailed account of the rather haphazard way in which the geologic column was originally developed,

> "...I wonder how many of us realize that the time scale was frozen in essentially its present form by 1840? A bit of western Europe, none too well, and a lesser fringe of North America. All of Asia, Africa, South America, and most of North America were virtually unknown. How dared the pioneers (of this theory) assume that their scale would fit the

rocks in these vast areas, by far most of the world? Only in dogmatic assumption--a mere extension of the kind of reasoning developed by Werner from the facts in his little district of Saxony. And in many parts of the world, notably India and South America, it does not fit. But even there it is applied! The followers of the founding fathers went forth across the earth and in Procrustean fashion made it fit the sections they found, even in places where the actual evidence literally proclaimed denial. So flexible and accommodating are the 'facts' of Geology."[12]

Then again, the radiometric scheme itself is afflicted with so many possibilities of error and conflicting dates that the system is totally dependent on previously agreed upon dates for any geochronometrical usefulness.

5. Finally, we must understand, that, as Jeletzky informs us, there are no physical characteristics which are authoritative in the dating of rocks,

> "The more than amply proved and almost unanimously recognized impossibility of establishing any practically useful broadly regional or worldwide geologic time scale based on the physical-stratigraphical criteria alone for the vast expanse of pre-Cambrian time, supplies conclusive proof that these phenomena are devoid of any generally recognizable geologic time significance."[13]

Actually the earth is dated primarily by use of the fossil record which is forged into a chronological sequence on paper that reflects the assumptions of evolutionary theory. In recent times, however, radiometric dating has been introduced as a means of guaranteeing an absolute value to the business of dating the earth. The modern use of this method, however, has found it to be seriously wanting as a reliable geochronometer.

Radiometric Dating

The greatest fault of all radiometric dating methods is the enormous importance which must necessarily be placed on assumptions which are unprovable and contrary to what we know about the nature of rocks in the real world. The first great assumption is that the radiometric decay chains have occurred within a closed system. If anything from within the system was deleted or if anything that was without the system was added over the years, then the system has been contaminated, and our measurements and readings have to be adjusted according to the degree and nature of alteration. However, we know of no real closed system in nature. Furthermore, the longer the time needed for the system to remain closed, the smaller the possibility of it doing so. For it to be closed for millions of years is an absurd assumption.

The second unprovable assumption is that the rocks contained no daughter products (lead, strontium, argon) at the beginning. If lead or any of the other daughter components such as strontium were present at the beginning, we must know in what proportions so that we can establish a ratio between the initial parent-daughter percentages of the past and the percentages detected in the rocks today. The problem is that there were no witnesses to the ratio of the parent/daughter components at the beginning. The methods that have been proposed for determining the original ratios are highly presumptive and impossible to confirm scientifically.

Obviously, it is impossible for anyone to know the initial components of such a system. Merely because it can be shown that there is in fact a radiometric decay process going on now does not make it possible for us to discern the initial radiometric state of the rocks. Henry Faul, who wrote one of the most important books on this subject, acknowledges the challenges that this assumption makes to the theory's credibility,

> "Two important assumptions are implicit in this equation: First, that we are dealing with a closed system. And second, that no atoms of the daughter were present in the system when it formed. These assumptions furnish the most serious limitations on the accumulation clock. Rigorously closed systems probably do not exist in nature, but surprisingly, many minerals and rocks satisfy the requirement well enough to be useful for nuclear age determination. The problem is one of judicious geologic selection."[14]

For all intents and purposes, Faul's use of the word "judicious" simply means that if the date obtained fits comfortably into the preferred evolutionary time unit, as expressed in the dates incorporated in the geologic column, which was decided upon between 1830 and 1840, then it is accepted. If it does not fit, then it is rejected.

The third assumption is that the process rate did not change over the millions or billions of years. However, many factors influence process rates in nature. If it is evident that the rates are subject to change now in the world we observe, how can we, with any certainty, demand that they did not change in the past? In the information below, the three faulty assumptions referenced above will be discussed in detail.

Uranium to Lead

The uranium method was the first to be used historically. Its importance is also appreciated when we realize that all of the other methods are calibrated against it. Of course, it may be a surprise to the man on the street to realize that this method is used primarily to date pre-Cambrian rock and not sedimentary rocks. Ancient life forms are now seen as fossils; and where are the fossils? In the sedimentary rocks. In the pre-Cambrian, fossils are almost non-existent. Fossils, as we shall see later, are really the main way that rocks are dated from an evolutionary standpoint. Since there are virtually no fossils in the Pre-Cambrian rock, the uranium lead method must therefore be employed. These rocks are supposed to be the earth's oldest rocks. Therefore, once the radiometric parameters are set with regard to these rocks, all other dating methods are obligated to restrict their dates to younger ages.

The uranium to lead decay chain is commonly referred to as the alpha decay process. Actually, three decay chains are present:

1. Uranium 238 decays into Lead 206 in addition to eight helium atoms. The half life is 4.5 billion years.
2. Uranium 235 decays into Lead 207 plus seven helium atoms. The half life is .07 billion years.
3. Thorium 232 decays into Lead 208 plus seven helium atoms. The half life is 14.1 billion years.

Many times a rock may also contain Lead 204 as well as these other radioactive minerals. Lead 204 does not appear to have any radioactive precursor.

Now let us see how the scientific evidence conflicts with these assumptions.

Closed System

1. Uranium minerals exist in "open" not "closed" systems. It is easy for uranium to be leached out by ground water. The intermediate element, radon gas, moves in and out of the system without any difficulty. One of the authorities in this field, Henry Faul (previously quoted with regard to the difficulties of this assumption) states,

> "Uranium and lead both migrate (in shales) in geologic time, and detailed analyses have shown that useful ages cannot be obtained with them. Similar difficulties prevail in attempts to date pitchblende veins. Here again much chemical activity is known to take place and widely diverging ages can be measured on samples from the same spot."[15]

J.D. Macdougall, in his *Scientific American* article, "Fission Track Dating," also comments on the unreliability of the "closed system" assumption,

> "...ground-water percolation can leach away a proportion of the uranium present in the rock crystals. The mobility of the uranium is such that as one part of a rock formation is being impoverished another part can be abnormally enriched. Such changes can also take place at relatively low temperatures."[16]

No wonder then that Faul has acknowledged,

> "... Rigorously closed systems probably do not exist in nature..."[17]

Lunar rocks might be expected to present a truer or purer perspective, but instead, they also assist in rendering this dating technique as inadequate. In an article in *Science News* regarding the dating of moon samples, Evelyn Driscoll writes,

> "If all of the age-dating methods (rubidium-strontium, uranium-lead, and potassium-argon) had yielded the same ages, the picture would be neat. But they haven't. The lead ages, for example, have been consistently older. This led Leon T. Silver, of the California Institute of Technology, to study the temperatures at which lead volatilized (vaporizes) and moves out of the lunar sample. Theoretically, this could happen on the moon and this volatilized lead would become

"parentless"--separated from its uranium parent. More lead (parentless lead added to the material) would yield older ages."[18]

Then again, we would ask: How many different ages can one moon be? Only one if the radiometric process was truly reliable. However, Driscoll continues,

> "Isotopic ages have been obtained for material from five landing sites on the moon--those of Apollo's 11, 12, 14, 15 and Luna 16; each site has a different age. But in a given site, the ages also vary.... Ideally, however, any one basaltic rock from a given site should yield the same isotopic age, regardless of the method used."[19]

Meanwhile, back on the earth, M.R. Klepper and D.G. Wyant, writing in the *U.S. Geological Survey Bulletin* in 1957, report,

> "As much as 90 percent of the total radioactive elements of some granites could be removed by leaching the granulated rock with weak acid...as much as 40 percent of the uranium in most fresh-appearing igneous rocks is readily leachable."[20]

Furthermore, the isotopic value of the lead can be changed when free neutrons in the rock are at times captured by the lead. Lead 206 can be converted into Lead 207 and Lead 207 into Lead 208 by free neutron capture.

Lead 208 usually constitutes over half of the lead present in any given lead deposit. Consequently, since we know Lead 208 can be developed by free neutron capture rather than radiometric decay, we can never be sure how in fact the Lead 208 was obtained in any given sample. To insist that it is there by radiometric process only, is simply wishful thinking and has nothing to do with scientific evidence.

This latter point is demonstrated by the research of Dr. Melvin Cook. He has analyzed two of the earth's most important uranium-bearing ores. Both in Katanga and in Canada, the ores contained no Lead 204 (hence no common lead) and little or no Thorium 232. However, they did contain large amounts of Lead 208. The Lead 208 could not have come from thorium decay nor even from common lead contamination. Its only possible source was Lead 207 by neutron capture. In order to clarify this, let us review our decay chains for a moment:

- Uranium 238 decays into Lead 206
- Uranium 235 decays into Lead 207
- Thorium 232 decays into Lead 208

Note that in the aforementioned Katanga and Canadian ores, there is little or no Thorium 232, yet there is plenty of Lead 208. How did it get there? The abundance of Lead 208 that is there could have not come from the small amount of Thorium 232, nor from common lead (204) contamination. The only option left is that it must have come from Lead 207 by means of neutron capture.

Dr. Cook concludes that it is therefore reasonable to presume that all of the radiogenic isotopes of lead found in the uranium-thorium systems can be accounted for by the free neutron capture process. These findings make it impossible to assert with any authority that Lead 208 in a given sample was solely the result of uranium decay. Consequently, this decay scheme is completely neutralized as a reliable geochronometer.

Constant Rate

Textbook Evolution

> "Rocks generally contain small amounts of radioactive material that can act as natural clocks. Atoms of the same element that have different numbers of neurons are called *isotopes. Radioactive isotopes* have nuclei that emit particles and energy at a constant rate regardless of surrounding conditions."[21] (Mead A. Allison, Arthur T. DeGaetano, & Jay M. Pasachoff, *Earth Science*, Holt, Rinehart, & Winston)

2. Contrary to popular accounts, the rates of uranium-lead decay systems are quite susceptible to change. The prominence of the free neutron capture mentioned above demonstrates that these elements are effected by cosmic influences. The reversal of the earth's magnetic field would effect the environment and furthermore, super novas are also considered to have been interruptive of the radiometric decay systems. Actually nothing in nature appears to be beyond the possibility of change. At the time of the

writing of his book, Dr. William S. Beck was with the Harvard Medical School. In *Modern Science and the Nature of Life*, he cautions,

> "When all is said and done, there seems to be evidence that even the 'laws of nature' are changing. Modern physics suggests the possibility that changes are taking place in the speed of light and in the rates of chemical reactions. In other words, the universe is changing, and it becomes hazardous to attempt calculations concerning the very remote past and the future."[22]

Let us emphasize that the millions of years needed to support the conventional theory of evolution are extremely remote.

The Director of Research of the Innovative Concepts Association, Fred Jueneman, states,

> "Being so close, the anisotropic neutrino flux of the super-explosion must have had the peculiar characteristic of resetting all our atomic clocks. This would knock our Carbon-14, Potassium-Argon, and Uranium-Lead dating measurements into a cocked hat! The age of prehistoric artifacts, the age of the earth, and that of the universe would be thrown into doubt."[23]

Ten years later, Jueneman reported on the most recent understanding of radiometric decay rates,

> "The age of our globe is presently thought to be some 4.5 billion years, based on radio decay rates of uranium and thorium. Such 'confirmation' may be short-lived, as nature is not to be discovered quite so easily. There has been, in recent years, the horrible realization that radio decay rates are not as constant as previously thought, nor are they immune to environmental influences. And this could mean that the atomic clocks are reset during some global disaster, and events which brought the Mesozoic to a close may not be 65 million years ago but, rather, within the age and memory of man."[24]

Though there is much more to say on the subject of the age of the earth, we must draw the reader's attention to the implications of Jueneman's statement. Mesozoic means dinosaurs; this is, in uniformitarian geology, the age in which they lived. The recognition of the fact that the radio decay rates are not as constant as the uniformitarian school has presumed, goes far to put the dinosaurs, and/or the Mesozoic Age, "...within the age and memory of man." May I refer the reader again, therefore, to the chapter in my book, *Dinosaurs*, entitled "Evidences of a World-Wide Flood from a Study of the Dinosaurs." (This Flood which created the Mesozoic layers in which dinosaurs are found has, with regard to "the memory of man" been referenced by more than 270 cultures of the world).

Parent/Daughter Ratio

Textbook Evolution

> "As an atom emits particles and energy, the atom changes into a different isotope of the same element or an isotope of a different element. Scientists measure the concentrations of the original radioactive isotope, or *parent isotope*, and of the newly formed isotopes, or *daughter isotopes*. Using the known decay rate, the scientists compare the proportions of the parent and daughter isotopes to determine the absolute age of the rock."[25] (Mead A. Allison, Arthur T. DeGaetano, & Jay M. Pasachoff, *Earth Science*, Holt, Rinehart, & Winston)

3. There is no way of knowing for certain that the daughter products were not present in the original formation of the rocks. Modern, volcanic rocks have illustrated this point forcefully. A British engineer, Sidney Clementson, has made detailed studies of a number of volcanic rocks. Reports on their uranium ages have been published in Soviet Geophysical Journals. He has shown conclusively that in every one of the cases, the uranium-lead ages were much older than the known age of the rock. Most of them gave ages of over a billion years, even though the volcano erupted in modern times. As a result of his studies, Clementson concludes,

> "Calculated ages give no indication whatever of the age of the host rock."[26]

When we realize that the daughter products occur naturally in the earth, judgments in regard to the percentage that occurred primarily by radiometric decay, become speculative to such an extent that the "absolute age" concept is rendered unsubstantial. Evelyn Driscoll explains,

"What complicates things for the uranium-lead method is that non-radiogenic lead 204, 206, 207 and 208 also exist naturally, and scientists are not sure what the ratios of non-radiogenic to radiogenic lead were early in the moon's history....The problem of how much lead was around to begin with still remains."[27]

If the argument is put forth that the uranium and lead isotopes came up in the lava simultaneously at the time of the eruption, that will be only to the detriment of the "millions of years" concept. We may safely say, that if under scientific investigation we can clearly see that the parent and the daughter product emerges contemporaneously as a natural part of the mantle forming process, then we should rightly conclude that the uranium-lead ratios have no validity as geochronometers. They simply reflect the manner in which igneous rock in general and radioactive isotopes in particular are developed.

The dictum of the uniformitarian ideology is that the present is the key to the past. If anyone wishes to be consistent in geology with that principle, then we must assume that the mantle forming processes of ancient history are not significantly different than those of today. If parent and daughter products show up together in the mantle forming process of today, we must assume that in the distant past the isotopic ratios must have developed in a similar manner.

The Helium Candle

Since it is impossible to determine whether or not all of the lead in a sample is there entirely by radioactive decay, it should then be obvious that the uranium-lead radioactive decay system cannot be a reliable geochronometer. However, it is of great interest to note that the by-product of this decay system, Helium 204 should be able to function much more reliably as a geochronometer since all Helium 204 in the rocks must be there by means of radiometric decay. Uranium 238 decays into Lead 206, note that our numbers have gone from 238 down to 206, this represents a loss of 32 mass units. The reason for this is because for every atom of Uranium 238 that decays into Lead 206, eight helium atoms are formed. Each of these are made up of four particles, thus, we have a loss of 32 mass units. This means that there should be at least 8 times as much Helium 204 in the rocks than there is Lead 206.

However, when this phenomena has been investigated it has been found that there is only enough Helium 204 in the rocks to render an age for the earth of less than 10,000 years. This has been an unpleasant finding for promoters of long ages. Of necessity, they have argued that the Helium 4 has escaped into the atmosphere. However, when this has been investigated it has been found that there is only enough Helium 4 in the atmosphere to give an age for the earth of approximately 40,000 years. Furthermore, as Fanale & Schaeffer of the Brookhaven National Laboratory have found the "escape out of the rocks" idea is a very weak argument,

> "Studies of the helium method (2) have shown that low ages based on helium, obtained on common rock-forming minerals, do not necessarily reflect diffusive loss of helium from the lattices of those minerals; under ideal conditions, some mineral lattices even appear to retain helium quantitatively for longer than 108 years."[28]

Consequently, they have had to go to a tertiary assumption, namely that the Helium 204 must also have escaped out of the atmosphere. This is an unreasonable assumption since it can be demonstrated that Helium 203, which is a lighter gas than Helium 204 is in abundance in the atmosphere. How could the heavier gas escape and the lighter gas remain? Further, how much Helium 204 must have escaped out of the atmosphere in order for the billions of years age scenario to be maintained? In order to keep the earth at an age of 4.5 billion years the present amount of Helium 204 in the atmosphere would have had to escape 100,000 times.

Rubidium-Strontium Dating

In the Uranium-Lead decay chain, we have an alpha decay process. In the Rubidium-Strontium, however, we have a beta decay process. In this case, Rubidium 87 decays into Strontium 87. Some authorities say the approximate half-life is from about 47 billion years, while others say up to 120 billion years. This method must be calibrated against the Uranium-Lead method. Among the problems with this method are the following:

1. Strontium 87 minerals can be penetrated by an influx of Rubidium 87

minerals within the same rock. Professor Gunter Faure at the Ohio State University and Professor James Powell at Oberlin College, Ohio, have also found the Rubidium-Strontium system to be an open one,

> "These results indicate that even total-rock systems may be open during metamorphism and may have their isotopic systems changed, making it impossible to determine their geologic age."[29]

2. The same neutron capture process which forms Lead 208 from Lead 207 can also form Strontium 87 from Strontium 86.
3. The factors which effect the rate of potassium decay and uranium decay also would inevitably influence rubidium decay.

4. It is not difficult for Rubidium 87 to be leached out of a Rubidium-Strontium system. Finally, in their article on the Ancient Lithosphere, Brooks, James, and Hart report on the discrepancies between Rb-Sr and the real ages of the rock,

> "One serious consequence of the mantle isochron model is that crystallization ages determined on basic igneous rocks by the Rb-Sr whole-rock technique can be greater than the true age by many hundreds of millions of years. This problem of inherited age is more serious for younger rocks, and there are well-documented instances of conflicts between stratigraphic age and Rb-Sr age in the literature."[30]

Now before we look at the third of the three most important radiometric decay methods, let us underscore the uselessness of the first two. The Chinese Geochemist, Dr. Y.F. Zheng, comments,

> "The Rb-Sr isochron method has been one of the most important approaches in isotopic geochronology. But some of the basic assumptions of the method are being questioned at the present time."[31]

When Michael Tweedie wrote his book, *The World of Dinosaurs*, he, like other authorities, has attempted to deal with the question of time. He is aware of radiometric methods and confident that the rates are stable. Nevertheless, he writes,

> "...by ascertaining the amount of uranium and the amount of the relevant isotope of lead present in a sample of rock, an estimate can be

made of the date at which the rock crystallized. The disadvantage of this method is that uranium and thorium minerals are rather rare, and that it can only be applied to igneous rocks."[32]

Before we continue with this quote, let us pause to reflect upon a few startling facts—or at least they should be startling to those who read the popular textbooks about the "certainty" of dating fossils. The first is that fossils themselves, especially dinosaur fossils, are never dated. British paleontologist, Michael Benton, states,

"Although we cannot usually date the fossils themselves, we can measure the age of the surrounding rocks and guess when the plants and animals lived."[33]

Alan Charig, of the British Museum, informs us that,

"As yet there is no radiometric method (that is, one based on radioactivity) for the direct absolute dating of dinosaurs."[34]

The only method we have of dating material from a once-living organism is that of radiocarbon. However, radiocarbon yields ages of only thousands of years, not millions. Consequently, those who are convinced of the uniformitarian "vast ages" concept will not use radiocarbon to date fossils because it does not give the dates they need for their evolutionary scenario.

The second startling fact is that even the sedimentary material in which the dinosaurs are found is not often dated. This is because the radiometric methods that yield long ages, as Tweedie confesses, are used only on igneous rocks,

"The sediments themselves cannot be dated because the radioactive history of each crystal grain in a sediment starts with the cooling and solidifying of the igneous rock from which it was ultimately derived by erosion."[35]

Tweedie then tells of the difficulty of working with the Uranium-Lead minerals in igneous rocks,

"If these are intruded into fossil-bearing sediments, or are lavas overlying them, then obviously the sediments are older; if the sediments overlie the igneous rock they are younger. But how much older or younger is often difficult to say, so uranium and thorium are not very suitable for dating the fossil record. The same objection applies to the breakdown of radioactive rubidium to form strontium."[36]

A similar conclusion has been reached by Michael Rampino, Professor of Biology at New York University, and the late Richard B. Stothers who was previously with the Goddard Institute for Space Studies. In their work with flood basalts, they found that the errors involved in the Rubidium-Strontium age-dating methods were too great for the method to be considered useful,

"Rubidium-strontium (Rb-Sr) ages were considered only as secondary methods of dating because of their commonly large associated errors."[37]

Potassium-Argon Method

Now, once and for all, we may clearly understand that with regard to fossils in general, and dinosaur fossils in particular, the vast ages concept is supported by unreasonable assumptions and not by substantial evidence--the Potassium-Argon method. As Tweedie explains,

"In recent years these methods have been largely superseded by the Potassium-Argon method."[38]

Potassium is found in igneous and in some sedimentary rocks. In this decay chain, Potassium 40 decays into Argon 40. The half-life is reputed to be 1.3 billion years. By the beta decay process, Argon 40 also decays into Calcium 40. Its reliability as a geochronometer is undermined by the following facts:

1. Like Rubidium-Strontium, it also must be calibrated by the Uranium-Lead dating method. The ratio which determines the amount of the decay product which become argon rather than calcium is unknown by a factor of up to fifty percent. Furthermore, the decay rate is still unsettled. Therefore, values of these constants are selected which correlate to

potassium dates within the uranium dates. As we have shown however, uranium dates are easily distorted by a number of variables.

2. This, too, is an open system. Argon 40 is a gas and is susceptible to migration in and out of potassium minerals. In an article in the *American Journal of Science,* J.F. Evernden and colleagues acknowledge,

> "Processes of rock alteration may render a volcanic rock useless for Potassium-Argon dating...We have analyzed several devitrified gases of known age, and all have yielded ages that are too young. Some gave virtually zero ages, although the geologic evidence suggested that devitrification took place shortly after the formation of a deposit."[39]

One of the clearest expressions of the possibilities of the Potassium-Argon method being influenced by physical and chemical changes has been expressed by Roger Lewin in his book, *Bones of Contention.* In this passage, he is explaining why Leakey chose to use the newly developed Argon 40/39 method over the conventional Potassium/Argon method,

> "If instead the heating is done gradually in steps, then the argon will be released gradually too, the first argon to be released will be that near the surface of the crystal and as the temperature rises, argon from deeper and deeper within the crystal is expelled. This means a series of age determinations is produced, essentially giving a profile entirely unmodified since first formed. All the age determinations will be the same, the spectrum will be flat. If however, the rock has been chemically or physically disturbed in some way over time, so that argon began to leak from the crystal lattice, the first dates obtained will be younger than the ones from the center of the crystal which may have lost no argon."[40]

Note here, the admission by Rampino and Stothers that the rocks can be externally, physically, and chemically altered. Research on flood basalts further clarifies the openness of this system,

> "Recent work on individual flood basalt episodes has shown that the long tail of young ages in many cases is the result of argon losses from altered samples, whereas the oldest ages may have been affected by initial incorporation of some excess argon. The total apparent age spread can reach tens of millions of years in some cases."[41]

As pointed out by Louis A. Rancitelli and David E. Fisher in their article, "Potassium-Argon Ages of Iron Meteorites," Potassium itself is also very mobile,

> "The Potassium-Argon ages of the meteorites investigated ranged from 5 x 109 years to 15.6 x 109 years. As much as 80% of the potassium in a small sample of an iron meteorite can be removed by distilled water in 4.5 hours."[42]

3. The decay rate of potassium is also subject to change. It could easily have been faster in the past than in the present. This, as in the case of the Uranium-Lead process, would be due to an increase in nutrino flux caused by past intermittent increases in cosmic radiation at the earth's surface.

4. It seems likely that Argon 40 is often incorporated into potassium at the time the original rock is being formed. As Henry Faul, one of the original researchers with regard to radiometric dating, reports, argon is common both to the atmosphere and to the crustal rocks,

> "It is difficult to be certain whether a significant amount of environmental argon was originally enclosed in any K-Ar system used for age determination because there is no basis for estimating the isotopic composition of original argon."[43]

There is an abundant amount of Argon 40 available, meaning that at least some of the Argon 40 in every potassium mineral was not solely from potassium decay, but was in part from the environment. As Laughlin, writing in the *Journal of Geophysical Research*, admits,

> "It is obvious that incorporation of excess Argon-40 may lead to significant errors in the age, particularly in the case of very young minerals or rocks, or of those with low potassium contents."[44]

This problem has been graphically illustrated by recent attempts to date the Hawaiian volcano Kilauea which erupted only about 200 years ago. Recently, attempts have been made to date these rocks by the Potassium-Argon method. The results demonstrate graphically, the unreliability of this method. In an article about deep ocean basalts, C.S. Noble and J.J. Naughton, writing in science, reveal,

"The radiogenic argon and helium contents of three basalts erupted into the deep ocean from an active volcano, Kilauea, have been measured. Ages calculated from these measurements, increase with sample depth up to twenty two million years for lavas deduced to be present. Caution is urged in applying dates from deep-ocean basalts in studies on ocean-floor spreading."[45]

In other words, the deeper the measurements were made, the greater the age they obtained for the volcanic rock.

Such experiences have clarified the fact that potassium-argon dating has some serious pitfalls, as Rampino and Stothers explain,

"Most flood basalts have been dated by the use of the 40K-40Ar method on whole-rock samples or mineral separates. However K-Ar ages are subject to a number of uncertainties arising from loss of radiogenic 40Ar due to slow leakage from poorly retentive, weathered, or hydrothermally altered minerals (5) or to complete loss during contact metamorphism around later intrusions. (6) Furthermore, any excess radiogenic argon incorporated into igneous rocks when the magmas crystallized (from xenoliths, for example) could result in apparent K-Ar ages that are older than the actual time of crystallization."[46]

At the time of the world wide flood, volcanic action was global. The consequences of the volcanic activity and the transference and deposition of sedimentary material at different locations all over the earth, can be graphically illustrated by the many sites where fossils have been found. These sites often contain not only sedimentary material (by which means the bones were turned into fossils) but also blankets of volcanic ash known as tuffs. One of the most vivid expressions of the unreliability of the radiometric dating methods involves a controversy over the dating of one of these tuffs in Africa.

During the 1970's a most significant squabble occurred among a variety of scientists regarding the dating of a volcanic tuff geologically associated with a theorized early homo and a collection of stone tools. This argument had to do with both the usefulness of certain advanced radiometric dating techniques and with the antiquity of man.

The specimens, a complete cranium of the Australopithecus Robustus type and a partial cranium of a supposed early homo were found by Richard Leakey near Lake Turkana. In association with Kay Behrensmeyer, a graduate student from Harvard, they found a variety of stone tools amidst volcanic ash. It was thought by Leakey and his associates, that the volcanic ash would be an excellent material for the application of the potassium-argon dating method.

Two experts, Jack Miller and Frank Fitch, were asked to apply the potassium-argon dating technique to the tuff found at this site. The controversial tuff came to be known as the KBS Tuff (Kay Behrensmeyer Site). Fitch informed Leakey that the samples sent to him (pumice and feldspar crystals) had yielded a date of 2.4 million years.

After receiving the first date, Lewin reports that it was decided to use an even more complex dating method developed by Miller and some other geochronologists. This method, called the argon-40/argon-39 method,

> "...would result in this tuff being incontrovertibly dated and with greater accuracy than any other site in Africa or elsewhere."[47]

After using the "incontrovertible" method of argon-40/argon-39, the date received from Fitch and Miller was that of 2.6 million years for this particular tuff. Leakey liked this date very much because it made the "human" artifacts in this site the oldest yet found. The importance of the ancient date was quadrupled three years later when the famous skull 1470 was found *below* the KBS tuff. This meant that the skull (considered homo) must be even older than the 2.6 million years of the material above it. Leakey advertised this as the oldest member of the genus homo ever discovered. It also served to put him high on the charts as a famous and successful paleoanthropologist.

Much controversy resulted from the date given by Fitch and Miller, which dated not only the KBS tuff and the stone tools, but also indirectly, the famous skull 1470. As a result, the two men tried again and again to apply their dating method to the various samples from the KBS tuff. However,

> "they never again obtained 2.61 from their experiments. For instance, at a conference in Nairobi held in September 1973 they presented 41

separate age determinations of the KBS tuff, which varied between 223 million and 0.91 million years."[48]

During the years of controversy that followed, Fitch and Miller held fast to their 2.61 million year age (later modified to 2.42 for technical reasons). With the battle over, the date was finally settled to the satisfaction of some and to the dissatisfaction of others,

"Each party had very good reasons for acting in the way it did. In addition, Leakey clearly had a vested interest in the older date, if for nothing else than because the claim for the oldest *Homo*, oldest stone tools, and so on was good for fund raising."[49]

As can be seen, many scientific reasons are not wholly responsible for the establishment of these absolute and "incontrovertible" dates. In fact there is no real objective science in the potassium-argon dating at all. In the final analysis, the potassium-argon method is a subjectively manipulated geochronometer whose multiple ages are selected to fit preconceived ideas. A. Hayatsu of the University of Western Ontario, states,

"In conventional interpretation of K-Ar age data, it is common to discard ages which are substantially too high or too low compared with the rest of the group, or with other available data such as the geological time scale. The discrepancies between the rejected and the accepted are arbitrarily attributed to excess or loss of argon."[50]

Like the theory of evolution, the potassium-argon method is a matter of preference, which is not supported by adequate scientific data. In the end, faith and imagination have as much to do with the process as do any physical facts. Evernden and John R. Richards acknowledge,

"Thus, if one believes that the derived ages in particular instances are in gross disagreement with established facts of field geology, he must conjure up geological processes that could cause anomalous or altered argon contents of the minerals."[51]

Another incident where potassium-argon (potassium 40) has been shown to be in error, has to do with the new attention being given the Deccan Traps. Here is a graphic demonstration of the fact that "ages" can be manipulated at will in order to achieve the goals of the researcher, and that potassium-argon is not a reliable geochronometer. With regard to the on-going mystery surrounding the dinosaur's extinction, two new theories have recently gained popularity in explaining the cause of this phenomenon. They may be described as the "whimper" and the "bang" theories. The former asserts that the time of their extinction was longer while the latter asserts that the time of their extinction was shorter. Both theories center around the iridium anomaly found in a thin clay layer and located at various boundary sites around the globe. It has been the desire of Courtillot and his colleagues to show that the Deccan volcanism is to be associated with the KT boundary and with the mass extinctions found in the fossil record. In this effort, they have labored to show that the iridium was not deposited instantaneously, but over a long period of time. The Deccan Traps had previously been dated by potassium-argon, and it had been accepted that they exhibit a duration of fifty million years (from eighty to thirty million years ago). This however was not a useful number to those who wanted to associate the Deccan Traps to the iridium anomaly in the KT boundary. In the four quotes below the French geophysicist, Vincent Courtillot, relates,

> "Until recently the lava samples form the Deccan Traps were thought to range in age from 80 million to 50 million years (estimated by measuring the decay of the radioactive isotope potassium 40 in rocks). Whether this range was real or just reflected an error in measurement was unknown."[52]

So, the age of the Deccan Traps was subjected to a new review. First, the study of magnetic reversals was applied to the problem,

> "Judging from the usual frequency of reversals, our results implied that the volcanism could not have lasted much more than one million years. If so, the spread of ages found by potassium 40 dating must have been wrong."[53]

Then other researchers applied the use of argon-argon dating,

> ". . .to determine how much potassium 40 had decayed during the lifetime of the rock samples. Their results confirmed that the Deccan flows were laid down over a relatively brief period. Age estimates for the Deccan lavas now cluster between 64 and 68 million years, and

much of the remaining scatter in ages may result from alteration of the samples or differing laboratory standards."[54]

As can be seen, this reduced the duration of the Deccan trap volcanism from 50 million years down to four million years. Finally, the use of fossil evidence, along with the study of magnetic reversals, achieved an even shorter age,

"Magnetic and fossil studies together reduce the estimated duration of Deccan volcanism to about 500,000 years, the best time resolution that can be obtained using present techniques."[55]

What all this shows, is that "ages" come and go according to the point that seems necessary to be made at the time. With so many millions and billions of years to play with, the theory of evolution can remain viable even though several million years are subtracted here and there.

In conclusion, we need to understand that prior to the development of radiometric dating the age of the earth was "guessed at" on the basis of the assumption that evolution had occurred. These "guesstimates" were then written up in a variety of periodicals as if they were scientifically accurate. Actually the age of a particular stratum was based upon the fossils that were found in it (in particular index fossils). John Reader in his book, *Missing Links*, provides us with valuable insight into the science of age dating prior to the development of radiometric dating. Regarding the discovery of Zing and the attempt to date it, Reader writes,

"Something more precise was called for—an actual time scale that would give ages in the earth rather than geological eras. Estimates of fossils' actual ages frequently appeared in books and articles, it is true, but they were very rough estimates based on guesses of how fast evolution proceeds, or how long sedimentary deposits take to accumulate. These estimates were always known to have little scientific value and they were produced primarily for popular publications (where their inconstancy must surely have confused as many as were enlightened)."[56]

By the time Zinganthropus was found, radiometric dating had been developed and was waiting in the wings to provide "precise" dates from then on. However, repeated attempts to apply this method have shown it to be unreliable, as we have previously illustrated from the Skull 1470 and KBS Tuff controversy.

The conflicts are found not merely to exist from rock to rock, but even within a single rock. Joan C. Engels, in the *Journal of Geology*, reveals,

> "It is now well-known that K-Ar ages obtained from different minerals in a single rock may be strikingly discordant."[57]

Little by little researchers are beginning to realize that absolute dates based on radiometric dating are not so absolute as previously believed. As E.H. McKee and D.C. Noble, writing in the *Geological Society of America Bulletin*, illustrate,

> "Perhaps the most questionable parts of geology are the dates, which are bound to be lacking in absolute proof."[58]

Without reference to the fossils and the "textbook" geologic column, which was established by "guesstimates", radiometric dating really has very little value. This fact has been openly acknowledged by, Derek Ager, the former president of the British Geological Association. He states that radiometric dating,

> "...is an exceedingly crude instrument with which to measure our strata and I can think of no occasion where it has been put to an immediate practical use. Apart from very "modern" examples, which are really archaeology, I can think of no cases of radioactive decay being used to date fossils....ever since William Smith at the beginning of the 19th century, fossils have been and still are the best and most accurate method of dating and correlating the rocks in which they occur."[59]

The discrepancies in fact, are becoming rather notorious. Even the *Penguin Dictionary of Geology* records,

> "It should be noted that dates (absolute dates) obtained by different methods (radioactive dating methods) commonly show some discrepancies... As the committee on the measurement of geological time said in 1950, 'These figures (i.e. dates) are, as railway timetables say, subject to change without notice.'"[60]

Professor Richard L. Mauger of East Carolina University exposes a major flaw of evolutionary dating in commenting that the dates which fit the evolutionary assumptions are accepted, while those which do not are rejected,

> "In general, dates in the 'correct ballpark' are assumed to be correct and are published, but those in disagreement with other data are seldom published nor are discrepancies fully explained."[61]

Now what, may we ask is this "other data?" Is it some scientifically substantive mechanism for firming up dates that are a bit shaky? Not at all, the "other data" is basically our old textbook column which was established by guesses and evolutionary assumptions long before the advent of radiometric dating. As Hayatsu explains in the *Canadian Journal of Earth Sciences*,

> "In conventional interpretation of K-Ar age data, it is common to discard ages which are substantially too high or too low compared with the rest of the group or with other available data such as the geological time scale. The discrepancies between the rejected and the accepted are arbitrarily attributed to excess or loss of argon."[62]

Again and again it is obvious that the textbook column remains the cannon with which all dating methods must comply.

We have previously referenced the difficulty in attempting to ascertain the quantity of argon that could be generated by radiometric decay (Faul, 1966), therefore the attempt to attribute an undesirable age "to excess or loss of argon" must be seen as a totally subjective decision. Nevertheless, by what means do the scientists resolve the continuing conflicts between the desired ages and the ages that are derived by radiometric dating? The "scientific" means is a matter of more subjective guesswork. In the comments of Evernden and Richards, their use of the word "conjure" is most apt,

> "...if one believes that the derived ages in particular instances are in gross disagreement with established facts of field geology, he must conjure up geological processes that could cause anomalous or altered argon contents of the minerals."[63]

The only absolutes left, therefore, in the many attempts to establish beyond doubt that the earth is millions and billions of years old have been candidly expressed by William Stansfield of Cal Poly in California in his science textbook,

> "It is obvious that radiometric techniques may not be the absolute dating methods that they are claimed to be. Age estimates on a given geological stratum by different radiometric methods are often quite different (sometimes by hundreds of millions of years). There is no absolutely reliable long-term radiological clock.'"[64]

One observer indicates that some astronomical observations also support this conclusion. Eddy has pointed out that there is no evidence based solely on observations that the sun is 4.5 x 100 years old,

> "I suspect," he said, "that the sun *is* 4.5 billion years old. However, given some new and unexpected results to the contrary, and some time for frantic recalculation and theoretical readjustment, I suspect that we could live with Bishop Usher's value for the age of the earth and sun. I don't think we have much in the way of observational evidence in astronomy to conflict with that."[65]

Now, since we know that there are a variety of decay schemes with half lives of varying rates, we need to ask the question, "Why were the particular decay systems of uranium-lead, rubidium-strontium and potassium-argon chosen?" This question is answered for us by Adrian E. Scheidegger in his book, *Principles of Geodynamics*,

> "The strontium decay scheme mentioned above is not only one of several nuclear decay-reactions occurring in nature which are suitable for radioactive age determinations. For a naturally decaying material to be suitable for such age determinations it is necessary, firstly, that its half-life be comparable with the age of the Earth..."[66]

This means that the age of the earth is not determined by radiometric dating. It was determined long before radiometric dating was invented. Radiometric dating is offered as a means of confirming (or to make more exact) the already assumed age of the earth. By what means then, was the age of the earth actually determined? The only answer is—by means of estimation. As mentioned earlier, evolution was first assumed to be a fact.

Then it was assumed that the slow and gradual processes which develop or break down the crustal layers of the earth today were equally responsible for the original development of the earth's crustal layers. To reiterate Reader,

> "'Estimates of fossils' actual ages frequently appeared in books and articles, it is true, but they were very rough estimates based on guesses of how fast evolution proceeds, or how long sedimentary deposits take to accumulate."[67]

Fossils Date the Rocks

At the beginning of this book, I mentioned that the primary means of establishing the vast ages of the geologic column was, by estimation, comparing sediments and fossils. Despite popular notions to the contrary, the geologic column does not get its dates from radiometric dating methods. The primary means of dating the rocks, and the fossils they contain, is by use of the fossils themselves. The *Encyclopedia Britannica* tells us,

> "In the first place, fossils are regarded as time-indices, on the general ground that any particular species comes into existence, lasts for a certain length of time and then dies out completely, never to recur."[68]

As E.M. Spieker also points out for the American Association of Petroleum Geologists,

> "And what essentially is this actual time-scale? On what criteria does it rest? When all is winnowed out and the grain reclaimed from the chaff, it is certain that the grain in the product is mainly the paleontologic record and highly likely that the physical evidence is the chaff."[69]

The late Professor Carl Pantin, a British zoologist, also states,

> "Now geologists are very interested in species—you recognize particular kinds of rocks by the particular species of fossils you find in them."[70]

Due to faith in the theory of evolution and the espoused biological sequence of organisms, the fossils, as related to that theorized biological sequence, have been extremely useful in the construction of the smallest possible time segments on the geologic column. With regard to the assumed biological sequence, Carl Dunbar of Yale acknowledges,

> "...we have assumed as an established fact the gradual evolution of life throughout geologic time. Fossils, thus provide us with a chronology 'on which events are arranged like pearls on a string.'"[71]

W.M. Elasser of the University of Maryland stresses the importance of the assumption of evolution to the construction of the geologic column,

> "As is well known, the order of the geological strata is fixed entirely by means of fossils; thus, the geological method presumes the existence in these periods of living beings of gradually increasing complexity."[72]

The glaring problem here is that the "gradually increasing complexity" theory, which justifies the ordering of the layers as presented in the textbooks' sections on the geologic column, remains an unproven theory. L. Harrison Matthews writing a new introduction to the *Origin of Species* remarks,

> "The fact of evolution is the backbone of biology, and biology is thus in the peculiar position of being a science founded on an unproven theory —is it then a science or a faith?"[73]

Though the geologic column is at times connected with the word "relative" with regard to ages, it should be abundantly clear that that word is to be taken lightly. For the geologic column, as constructed by the use of fossils, has, over the years, been accredited by evolutionists with an amazing degree of geo-chronometric accuracy.

Fortey, a paleontologist at the British Museum of Natural History, goes into much detail illustrating the importance of fossils for dating the rocks. Below are a few of his comments, which should serve to illustrate this point,

1. "The finest subdivision became (and remains) that of the *Zone*. A zone is a division of geological time characterized by a particular assemblage of fossil species."

2. "The kinds of fossil organisms used as the basis for zones varies through geological time, as one group rises to prominence only to be replaced by another."

So, for all of the popular propaganda, it is to be noted that the paleontologist with his fossils, and not the geo-chronometrist with his radiometric dating devices, makes the most important contribution to the business of dating the earth.

3. "Radiometric ages always carry a margin of error (+/-) of some magnitude, and certainly more than the length of time for a fossil zone. So the relative time scale provided from fossil evidence is more accurate, and the use of fossils is not likely to be displaced by radiometric ages."[74]

Fortey's enunciation and delineation of the separate divisions of the geologic column is still entirely theoretical. Fifteen years earlier, Adolph Knopf of Yale University had stated very clearly the difficulties burdening this theory,

> "An urgent task for geology is to determine, in years, the length of the eras, periods, and 'ages' (time spans of the stages) and, eventually of the zones. Not a single one of them--eras, periods, and ages, let alone zones--has yet been reliably determined. This statement is possibly surprising in view of the fact that almost any modern writer can produce a geologic timetable that gives precise datings and lengths of eras and systems and even of some of the smaller subdivisions...All other absolute ages have been derived from the three radioactive tie points by interpolation based on thickness of strata or by 'reasoned guesses'...."[75]

As Carl Dunbar explains, the fossils which are "more useful" in dating the rocks are known as index fossils,

> "Inasmuch as life has evolved gradually, changing from age to age, the rocks of each geologic age bear distinctive types of fossils unlike those of any other age. Conversely, each kind of fossil is an *index* or *guide fossil* to some definite geologic time....During the last hundred years, paleontologists in many parts of the world have co-operated in

gathering such a mass of this kind of information that it is now as easy for a trained specialist to identify the relative geologic age of a fossiliferous rock formation as it is to determine the relative place of a sheet in a manuscript by its pagination. Fossils thus make it possible to recognize rocks of the same age in different parts of the Earth and in this way, to correlate events and work out the history of the Earth as a whole. They furnish us with a *chronology,* 'on which events are arranged like pearls on a string.'"[76]

Even though the discovery of an index fossil makes the dating automatic, it is acknowledged that there are problems. Stansfield writes,

"Certain fossils appear to be restricted to rocks of a relatively limited geological age span. These are called index fossils. Whenever a rock is found bearing such a fossil, its approximate age is automatically established....This method is not foolproof. Occasionally an organism, previously thought to be extinct, is found to be extant. Such 'living fossils' obviously cannot function as index fossils except within the broader time span of their known existence."[77]

In fact, this method has more severe problems than the aforementioned, which have to be continually explained away. Evolution theorists firmly deny that all of these fossils represent organisms which lived at the same time. According to evolution these various organisms lived in successive ages. Theoretically then, as George McCready Price challenged many years ago, since the index fossils are supposed to be

"'...restricted to rocks of a relatively limited geological age-span' (Stansfield, 1977), and '...unlike those of any other age' (Dunbar, 1949), they cannot possibly show up in a succeeding age. Index fossils cannot enjoy a cyclical experience. However this has occurred many times. For instance, In Scotland five such repetitions of key fossils were reported over half a century ago, thus initiating the famous 'Highland Controversy'. In the Alps six repetitions have been reported. In other places the number of such major repetitions varies...."[78]

Another deficiency in this method is the fact that in order to get the fossils to identify "ages," circular reasoning must be employed. J.E. O'Rourke, writing in the American Journal of Science, says,

"The intelligent layman has long suspected circular reasoning in the use of rocks to date fossils and fossils to date rocks. The geologist has never bothered to think of a good reply, feeling that explanations are not worth the trouble as long as the work brings results. This is supposed to be hard-headed pragmatism."[79]

R.H. Rastall, writing for the *Encyclopedia Britannica*, also acknowledges this problem,

"It cannot be denied that from a strictly philosophical standpoint geologists are here arguing in a circle. The succession of organisms has been determined by a study of their remains embedded in the rocks, and the relative ages of the rocks are determined by the remains of organisms that they contain."[80]

Niles Eldredge, who championed Ramapithicus for many years as an evolutionary ancestor, sees this circularity phenomenon as something of a contradiction to the evolutionary modus operindi,

"And this poses something of a problem: if we date the rocks by their fossils, how can we then turn around and talk about patterns of evolutionary change through time in the fossil record?"[81]

The straw-that-breaks-the-camel's-back in this matter, is the recognition of the fact that the whole thing only works if it is assumed that evolution has taken place. So then, the first premise of this argument is the assumption of evolution. The second premise, is the assumption that it took vast ages. With those assumptions accepted, then the theorist can go on to date the rocks by the fossils. Addressing the Geological Society of America, Dr. Hollis Hedberg firmly ties the use of fossils as time indicators to the assumption that evolution is a fact of science,

"That our present-day knowledge of the sequence of strata in the earth's crust is in major part due to the evidence supplied by fossils is a truism. Merely in their role as distinctive rock constituents, fossils have furnished, through their record of the evolution of life on this planet, an amazingly effective key to the relative positioning of strata in widely separated regions and from continent to continent."[82]

Note the phrase, "...through their record of the evolution of life on this planet..." In other words, faith in the concept of evolution makes it possible for the fossils to be used as age indicators.

However, as O'Rourke confesses,

"The procession of life was never witnessed, it is inferred."[83]

Evernden and associates also confirm the necessity of first assuming evolution to be true before the dating game can work,

> "Vertebrate paleontologists have relied upon 'stage-of-evolution' as the criterion for determining the chronological relationships of faunas. Before establishment of physical dates, evolutionary progression was the best method for dating fossiliferous strata."[84]

Both paleontologists and geologists rely on the concept of organic evolution (believed) for establishing and identifying "ages". As von Engeln and Caster relate in their book, *Geology*, historic geology relies chiefly on paleontology, the study of fossil organisms,

> "The geologist utilizes knowledge of organic evolution, as preserved in the fossil record, to identify and correlate the lithic records of ancient time."[85]

Nevertheless, the theory of uniformitarianism, (even with the recently permitted interpolations of catastrophe) is still in conflict with the observable facts of geology. A.C.M. Laing, in a letter to the editor of *The Australian Geologist*, states,

> "For nearly 30 years working with recent geological graduates, I have to teach each one to forget the theories he was taught, just observe what is actually there and record it."[86]

In a word, there is no real objectivity in the business of dating the rocks by the fossils or in the use of radiometric dating. How then do we date the earth? First, one must believe in the theory of evolution and assume that animals evolved from the simple to the complex over millions of years of time. Secondly, then the fossil organisms must be placed in a theorized vertical sequence from simple to complex. Then, estimates are offered for the length of time expected for each theorized geological era. Finally, when these smaller and greater eras are established, radiometric dating is used to provide a more "precise date" within the already firmly established estimated dates. Everything is dependent upon the theorized placement of the fossils and upon the assumption that evolution really happened. In the long run, nothing with regard to the age of the earth is dependent upon radiometric dating. Ager summarizes it very nicely,

> "...fossils have been and still are the best and most accurate method of dating and correlating the rocks in which they occur....I can think of no cases of radioactive decay being used to date fossils."[87]

In the final analysis, it is clear then, that radiometric dating methods are not reliable, and that the entire geologic column as viewed in the textbooks was developed by the use of estimation on the one hand, and by faith in the concept of organic evolution on the other. The theory of evolution demands vast ages, which radiometric dating and the unproven assumptions of evolution and uniformitarianism cannot provide.

Archaeological Dating vs. Evolutionary Paleontological Dating

In dealing with the pseudo-science associated with evolutionary dating it will be useful to start out with an everyday experience that people can have on the Amtrak train. If you are sitting on the left side of the train facing forward and you should fall asleep for several minutes, you will occasionally wake up and look out the left window to see a train whizzing by very fast. The immediate challenge to your mind is, 'is my train standing still and the train in the window rushing past us?' or 'is that train standing still and are we rushing past it?' If you keep looking out the left window there is no way to tell. You have to have a third reference point. So you look out the opposite window to your right. If you see a telephone post standing still, you know that your train is not moving. This means that the other train is the one moving. If you see the telephone poles whizzing by you on the right, then you know for sure that your train is actually moving. So too, in attempting to date the sedimentary layers of the earth (paleontology/geology) or strata layers (archaeology), there must be three reference points in order to have any degree of accuracy.

Over the years archaeologists have become increasingly accurate with regard to dating their finds because they have established three reliable reference points. In paleontology, however, with regard to dating the fossils or the sedimentary layers, it will be shown that they do not have three factual reference points. The book, *Digging up the Bible*, by Moshe Pearlman will provide an excellent literary source for illustrating the real science involved in archaeological dating.

The man who discovered the means by which strata levels could be accurately dated in the world of archaeology was Sir Flinders Petrie. Over time, as he surveyed the ancient ruins which he came across, he began to realize that potsherds could be of great value in dating the various layers of strata,

> "The vessels themselves had a short life, since they broke easily and could not be mended. But the material, though brittle and breakable, was itself imperishable, neither rotting nor decomposing...."[88]

This means that these broken pots could stay in the ground for many, many years and still be recognizable and useful for dating purposes. Further, Petrie noticed that,

> "Not all the pottery he had encountered was the same. There were differences in shape, color, type and location of handle, style of decoration, combination of kinds of clay, technique of manufacture (hand or wheel) and method of firing."[89]

Since these pots were made of material that was easily broken, the potters were constantly having to make new ones. As the potters did this generation after generation they endeavored to make improvements. This resulted in different colorations, styles, shapes, etc.

> "Thus, every decade or generation, and certainly every century, would produce its own style, and each style would be specific to its particular period. These periods would be brief, nothing like the ages of antiquity covering vast stretches of time....A pottery vessel might thus be given a date of origin to within a few years - which in archaeology is virtually pin-pointing in time."[90]

Some artifacts such as stone bowls or a wall could have been used for many generations, but a pottery style would have represented a much shorter period of time. Consequently, it would be,

> "...feasible to make the general presumption...that the remains of an ancient city in which such pottery was found belonged to that period."[91]

> "'If,' said Petrie, 'the varied pottery styles found on an archaeological site could be classified and correlated with their corresponding periods, not only would the history of the site be known but it was also likely that the classification would apply to all the sites in the region. It would therefore be possible to establish a standard chronology for the relative dating of all levels of settlement based on the pottery objects found in each stratum.'"[92]

This is exactly, as H.D. Hedberg confirms, what the evolutionary paleontologists attempt to do with the fossils,

> "However, going far beyond this, fossils have furnished, through their record of the evolution of life on this planet, an amazingly effective key to the relative positioning of strata in widely separated regions and from continent to continent."[93]

This statement shows that the evolutionists assume that if a fossil had been accurately reckoned to have lived during a certain geological period in the dim distant past, then whenever they found such a fossil in another area of the world, they could be sure that this particular sedimentary layer represented the same period of geological time as the sedimentary layer where the fossil was first found. The key phrase in Hedberg's statement is "through their record of the evolution of life". However, let's be clear, the main point of the search for fossils by the evolutionists is not just to learn of the animals that lived in the bygone past, it is to find transitional fossils that can demonstrate that the theory of evolution is indeed a fact. Unfortunately for the evolutionists, no such transitional fossils have been found, and this is readily admitted by many scientists. Further, the fossil record is so bad with regard to this goal that some have instituted an entirely new interpretive scenario called "punctuated equilibrianism", which attempts to scientifically confirm evolution as a fact on the basis of the lack of transitional fossils. In this they say that evolution happened, but it happened in spurts, which in geological time, were so short that no transitional fossils were ever fossilized.

So, in Hedberg's statements regarding the use of fossils, he is tacitly elevating the theory of evolution to an established fact. There are indeed vast quantities of fossils over the earth, and indeed one can find the same fossil species in North America as you find in Africa or in Asia. But, they are fully developed species. None of them are transitional species, which means that in such a statement as his, he is using the fully developed species as a time chronometer without any justification for doing so because there are no transitional fossils found to justify such confidence.

So, the fossils, they say, accomplish the same goal as does the pottery, according to Hedberg's statements. But how do they actually do this? They do this "through their record of the evolution of life." In other words, they date the stratum by elevating the theory of evolution to the level of a scientific fact.

Here again, as referenced below in a comment by Carl Dunbar, the key word that makes all of this evolutionary dating possible is "assumption,"

> "...we have assumed as an established fact the gradual evolution of life throughout geologic time. Fossils, thus provide us with a chronology 'on which events are arranged like pearls on a string.'"[94]

We will return to these quotes shortly and show that the "gradual evolution of life throughout geologic time" is not by any means "an established fact."

However, in the case of archaeological dating, these scientists are not required to be dependent on an assumption for their third reference point. They have, instead, an artifact which provides an objective date for the potsherd and thus the stratum. Moshe Pearlman continues,

> "Suppose, for example, the remains of an ancient city could be specifically dated by the rare discovery of an inscription, yet the pottery found therein was unfamiliar and missing from the list in the classification Petrie hoped would one day be compiled. Finding it in that stratum would give it the date. This pottery would then in its turn make possible the dating of other sites where it appeared and where other dating clues were absent."[95]

At Tel el Hesse, Petrie came upon a situation where there was a sixty foot high section of strata, which represented about 1,000 years of settlements. Here, he was able to put his pottery ideas to the test,

> "For the next six weeks, Petrie made vertical sections, and carefully noted the exact levels at which the various styles of pottery lay. The results soon confirmed his theory of the direct correlation between pottery and stratigraphy, namely, the chronological sequence of the strata in an ancient site; for he was able to show that each period had pottery which was characteristic of that period alone, and distinguishable by the trained archaeologist from corresponding articles of pottery belonging to earlier and later periods."[96]

Because of the uniqueness of the stratigraphy at Tel el Hesse and because of his wise interpretation of the use of pottery, this new method of dating was successfully begun at Tel el Hesse. However, it was left to the eminent archaeologist, William Foxwell Albright, to finally bring this concept into prominence as a truly reliable dating methodology,

> "Albright's excavation methods were punctilious. The remains he unearthed were notable. However the prime importance of his Tel Beit Mirsim dig was its establishment of pottery sequences on a firm basis. For the first time, the pottery was examined with scrupulous care, special attention being paid to its shape, texture, decoration and method of manufacture, and to the archaeological context in which it was found. By the precise differentiation between the pottery types, and each type related to a settlement level, it was possible to prepare a chronological sequence which could become a standard of comparison for the finds at all archaeological sites in the country. And this was done. Thereafter, Albright's ceramic typology system, with continued refinements of his own and others, became the archaeological key to the chronology, and therefore to the history, of every excavated biblical site."[97]

So then, the pottery dating method has been firmly established and has proven very reliable with regard to dating the various stratum in which the ruins of various ancient civilizations have been found. However, potsherds do not have dates on them. In order for these potsherds to be able to date the stratum, there had to be a third reference point, one that did provide reliable dates,

"The primary archaeological clues to the precise dating of biblical ruins are ancient writings, coins, and seals. Least rare are the seals, and particularly seal impressions. They have been found at a number of archaeological sites in Palestine, for they were much used in the entire region for sealing official documents, for marking personal objects belonging to notables, and for sealing container jars carried on the international trade routes which passed through Palestine."[98]

Though they are challenging to find,

"...ancient writings are of course the best evidence for absolute dating; the ideal which every archaeologist yearns to attain is a dedicatory inscription recording the date when a monument or structure was erected."[99]

A good example of the kind of finds with writing that can truly pinpoint a date for a particular stratum is as follows,

"Sixth-century pottery, in particular two jar handles stamped in Hebrew with the seal of 'Elyakim, the servant of Yochan', found amidst ashes in the latest stratum, point to the destruction of the last city of Debir during the Babylonian campaign of 589-587 BC."[100]

Now in review, let us consider the differences between the two systems of dating: radiometric dating, as used by the evolutionist, and pottery dating, as used by the archaeologist. In archaeological dating we learn that there are three reference points needed to resolve the problem of dating ancient strata. Recall my example above of the passing train and how to tell whether it was standing still or moving. The three reference points were the train that you were in, the train that appeared to be passing by, and finally and most importantly, the scenery that you saw when looking out the opposite window. Were those trees or telephone posts moving or standing still? Now, with regard to archaeological dating, we indeed have three objective reference points: the stratum, the potsherds, and the ancient inscriptions or perhaps a dated coin. By contrast, in evolutionary paleontological dating we have the following: the strata, the fossils in the strata, and...and...and a theory--the theory of evolution. This is plainly acknowledged by evolutionary scientists as referenced earlier,

"...we have assumed as an established fact the gradual evolution of life throughout geologic time. Fossils, thus provide us with a chronology 'on which events are arranged like pearls on a string.'"[101]

Unfortunately for the writer of this quote, Carl Dunbar, "the gradual evolution of life throughout geologic time" though it is assumed as established fact, it is in reality, not by any means an established fact. When Darwin's book occasionally gets reprinted they often ask a highly respected scientist to write the introduction. L. Harrison Matthews wrote the introduction to Darwin's *Origin of Species* as follows,

"The fact of evolution is the backbone of biology, and biology is thus in the peculiar position of being a science founded on an unproved theory...is it then a science or a faith?"[102]

There is no evidence that evolution is occurring in the present.

"Evolution, at least in the sense that Darwin speaks of it, cannot be detected within the lifetime of a single observer."[103] (David Kits)

There is no evidence that evolution occurred in the past.

"Evolution requires intermediate forms between species and paleontology does not provide them." (David Kits)

Darwin did not show how a new species develops in his, *On the Origin of Species.*

"Darwin never really did discuss the origin of species in his *On the Origin of Species.*"[104] (Ernst Mayr)

"One of the ironies of the history of biology is that Darwin did not really explain the origin of new species."[105] (Douglas Futuyma)

As yet, no scientist have been able to produce a new species by any means.

"No one has ever produced a species by mechanisms of natural selection. No one has gotten near it..."[106] (Colin Patterson)

The science community is still in the dark as to how speciation actually occurs.

> "Evolution is...troubled from within by the troubling complexities of genetic and developmental mechanisms and new questions about the central mystery--speciation itself."[107] (Keith S. Thompson)

Once a potsherd has been associated with an objective time dating artifact, such as an ancient inscription, that particular potsherd style can therefore be used to date stratum in a wide variety of geographical locations. This is precisely what the evolutionists try to do by means of their interpretation of the fossils. Again, Hedberg clarifies this point,

> "However, going far beyond this, fossils have furnished, through their record of the evolution of life on this planet, an amazingly effective key to the relative positioning of strata in widely separated regions and from continent to continent."[108]

Yes, this would work very well if it could be firmly and scientifically established that any particular fossil was in strata that was actually 350 million years old and that it represented an organism that actually evolved in so many million years from a less complex organism. However, this is not the case. Among the many commentators casting doubt on this idea is Henry Gee, the chief science writer for *Nature* magazine. In his article, "In Search of Deep Time," he states,

> "To take a line of fossils and claim that they represent a lineage is not a scientific hypothesis that can be tested, but an assertion that carries the same validity as a bedtime story--amusing, perhaps even instructive, but not scientific."[109]

These vast periods of time are based on the theory of uniformitarianism. In this view there has never been, on the earth, any geological processes other than what are currently happening, that is slow and gradual erosion and/or local floods. This view was developed for the purpose of denying that Noah's flood was global and that it was a flood of judgment. Observations, however, make it abundantly clear that if all we have ever had on the earth was slow and gradual erosion of sediments and local floods, there never could be a fossil record. Fossils can only be created by means of rapid deposition. So, with regard to time, notice again further remarks by Henry Gee,

> "The intervals of time that separate fossils are so huge that we cannot say anything definite about their possible connection through ancestry and descent."[110]

In summary, the two dating systems referenced above represent a dating system based on objective evidence and a dating system based entirely on subjective evidence. The archaeological dating is as follows:

The stratum is dated by a particular type of potsherd, which in turn is dated by an inscription or a coin (an objective, invisible time indicator). Whereas, in evolutionary dating (paleontological/radiometric) the stratum is dated by a fossil, which is dated by the theory of evolution. One must first reject the Genesis account of Creation and the tremendous evidence, including intelligent design, in its favor. Then, one must reject the Genesis account of the global flood and the tremendous evidence for world-wide catastrophism, as well as the more than 500 accounts of a global flood from the various cultures of the world. Finally, one must assume, by faith, that one small creature evolved into another by totally naturalistic means, for which there are no mechanisms for doing this, and one must assume that the sedimentary layers actually developed slowly and gradually at the rate of approximately 5,000 years per foot. If this would have been the case, there could not have ever been a fossil record.

Dunbar of Yale told us that, "...fossils provide the only historical, documentary evidence that life has evolved from simpler to more and more complex forms."[111] This means that if there are no transitional fossils found then there is no historical documentation for evolution. Steven Stanley of Johns Hopkins University, along with others, tell us, "Established species are evolving so slowly that major transitions between genera and higher taxa must be occurring within small, rapidly evolving populations that leave no legible fossil record."[112]

No "legible fossil record" means no "historical, documentary evidence" for evolution. After all is said and done, it remains that the main evidence for the theory of evolution is the theory of evolution.

The Genesis record remains the only explanation for the origin of the material world and of living things on the earth that is compatible with the laws of science, the laws of logic, and the observable facts of nature. By contrast, the modern theory of evolution is not compatible with the laws of science, the laws of logic, or the observable facts of nature. None of the facts of nature are in conflict with the Creation account as found in the book of Genesis. However, those who hold to the theory of evolution must do so in spite of the evidence, not because of it. In Lewontin's statement below, the word "science" must be understood to mean "evolutionary science." Science is what you see, and we do not see transitional animals between species today. There are no transitional fossils in the fossil record, and there are no unequivocal positive mutations.

> "We take the side of science *in spite* of the patent absurdity of some of its constructs, *in spite* of its failure to fulfill many of its extravagant promises of health and life, *in spite* of the tolerance of the scientific community for unsubstantiated just-so stories, because we have a prior commitment, a commitment to materialism."[113]

In light of the above comment, it is no wonder that scholars, such as Jerry Coyne of the Department of Ecology and Evolution at the University of Chicago, report as follows,

> "We conclude--unexpectedly--that there is little evidence for the neo-Darwinian view: its theoretical foundations and the experimental evidence supporting it are weak."[114]

Chapter Seven End Notes

---

[1] Lawrence Badash, *Scientific American*, Aug., 1989, p. 90
[2] *Earth Science: Geology, the Environment, and the Universe*, McGraw Hill/Glencoe, 2005, p. 554
[3] *Ibid.*, p. 556
[4] *California Life Science*, Pearson/Prentice Hall, 2008, p. 281
[5] John Reader, *Missing Links*, p. 157
[6] J.A. Jeletzsky, "Paleontology, Basis of Practical Geochronology." *Bulletin American Association of Petrolium Geologists,* volume 40, 4/1956, p. 685
[7] Edmund M. Spieker, "Mountain-Building Chronology and the Nature of the Geologic Time Scale." *Bulletin American Association of Petroleum Geologists,* (volume 40, 8/1956), p. 1805
[8] *Earth Science: Geology, the Environment, and the Universe*, McGraw Hill/Glencoe, 2005, p. 558
[9] *Ibid.*
[10] *Ibid.*, p. 562
[11] *Ibid.*
[12] Op Cit. Spieker, p. 1803
[13] J.A. Jeletzsky, op cit., page 684
[14] Henry Faul, *Ages of Rocks, Planets and Stars*, McGraw-Hill Book Company, Inc., 1966, page vi
[15] Ibid., page 61
[16] J.D. Macdougall, "Fission Track Dating," *Scientific American*, Vol. 235 (6):118
[17] Henry Faul, op cit., page vi
[18] Evelyn Driscoll, "Dating of Moon Samples: Pitfalls and Paradoxes," *Science News,* (volume 101, 1/1, 1972), p. 12
[19] *Ibid.*
[20] M.R. Klepper and D.G. Wyant. "Notes of the Geology of Uranium," *U.S. Geological Survey Bulletin,* No. 1046-F, 1957, p. 93
[21] Mead A. Allison, Arthur T. DeGaetano, & Jay M. Pasachoff, *Earth Science*, Holt, Rinehart, & Winston, 2008, p. 193
[22] William S. Beck, *Modern Science and the Nature of Life,* Harcourt, Brace, 1957, p. 170
[23] Frederick Jueneman, "Scientific Speculation," *Industrial Research,* 9/1972, p. 15
[24] Frederick Jueneman, "Secular Catastrophism," *Industrial Research and Development,* June 1982, p. 21
[25] Mead A. Allison, Arthur T. DeGaetano, & Jay M. Pasachoff, Op cit.
[26] S.P. Clementson, "A Critical Examination of Radioactive Dating of Rocks," *Creation Research Quarterly,* Volume 7, 12/1970, pp. 137-141
[27] Evelyn Driscoll, "Dating of Moon Samples: Pitfalls and Paradoxes," *Science News,* Volume 101, 1/1, 1972, p. 12
[28] Fanale & Schaeffer, *Science*, Vol. 149, p. 312
[29] Gunter Faure & James L. Powell. *Strontium Isotope Seology,* Springer-Verlag, 1972, p. 102
[30] C. Brooks, D.E. James, & S.R. Hart, "Ancient Lithosphere: Its Role in Young Continental Volcanism," *Science,* Volume 193, 9/17, 1976, p. 1093
[31] Y.F. Zheng, "Influences of the Nature of the Initial Rb-Sr System on Isochron Validity," *Chemical Geology - Isotope Geoscience* Section, vol. 80, 1989, p. 1
[32] Michael Tweedie, *The World of Dinosaurs,* Treasure Press, 1977, p. 18
[33] Michael Benton, *The Dinosaur Encyclopedia,* Simon and Schuster, Inc., 1984, p. 17
[34] Alan Charig, *A New Look at the Dinosaurs*, Mayflower Books, 1979, p. 36
[35] Op. cit., p. 18
[36] Op. cit., p. 18
[37] Michael Rampino & Richard B. Stothers, "Flood Basalt Volcanism During the Past 250 Million Years," *Science,* Volume 241, 8/1988, p. 664
[38] Op. cit., p. 18

[39] J.F. Evernden, D. E. Savage, G. H. Curtis, & G. T. James, "K/A Dates and the Cenozoic Mammalian Chronology of North America," *American Journal of Science,* Volume 262, 2/1964, p. 154
[40] Roger Lewin, *Bones of Contention*, Simon & Schuster, Inc., 1987, p. 193
[41] Michael Rampino & Richard B. Stothers, "Flood Basalt Volcanism During the Past 250 Million Years," *Science,* Volume 241, 8/1988, p. 664
[42] Louis A. Rancitelli, & David E. Fisher, "Potassium-Argon Ages of Iron Meteorites," *Planetary Science Abstracts, 48th Annual Meeting of the American Geophysical Union,* 1967, p. 167
[43] Henry Faul, *Ages of Rocks, Planets and Stars,* McGraw Hill Book Company, Inc., 1966, p. 32
[44] A. William Laughlin, "Excess Radiogenic Argon in Pegmatite Minerals," *Journal of Geophysical Research,* Volume 74, 12/15, 1969, p. 6684
[45] C.S. Noble, and J.J. Naughton, "Deep-ocean Basalts: Inert Gas Content and Uncertainties in Age Dating," *Science,* Volume 162, 10/11, 1968, p. 265
[46] Michael R. Rampino, and Richard B. Stothers, "Flood Basalt Volcanism During the Past 250 Million Years," *Science,* Volume 241, 8/3 1988, p. 663
[47] Roger Lewin, *Bones of Contention,* Simon and Schuster, 1987, p. 92
[48] *Ibid.*, p. 194-195
[49] *Ibid.*
[50] A. Hayatsu, "K-Ar Isochron Age of the North Mountain Basalt, Nova Scotia," *Canadian Journal of Earth Sciences,* Volume 16, 1979, p. 974
[51] J.F. Evernden, and John R. Richards, "Potassium-argon Ages in Eastern Australia," *Journal of the Geological Society of Australia,* Volume 9(1), 1962, p. 3
[52] Vincent E. Courtillot, "A Volcanic Eruption," *Scientific American,* 10/90, p. 85
[53] *Ibid.* p. 87
[54] *Ibid.*
[55] *Ibid.*
[56] John Reader, *Missing Links,* Little, Brown and Company, 1981, p. 157
[57] Joan C. Engels, "Effects of Sample Purity on Discordant Mineral Ages Found in K-Ar Dating," *Journal of Geology,* Volume 79, 9/71, p. 609
[58] E.H. McKee, and D.C. Noble, "Age of the Cardenas Lauas, Grand Canyon, Arizona," *Geological Society of America Bulletin,* 8/76, pp. 1188-1190
[59] Derek Ager, "Fossil Frustrations," *New Scientist,* Volume 100, No. 1383, 11/10, 1983, p. 425
[60] D.G.A. Whitten, and J. R. V. Brooks, *The Penguin Dictionary of Geology,* Penguin Books, 1972, p. 378
[61] Richard L. Mauger, "K-Ar Ages of Biotites from Tuffs in Eocene Rocks of the Green River, Washakie, and Vinta Basins, Utah, Wyoming and Colorado," *Contributions to Geology,* University of Wyoming, Volume 15(1), 1977, p. 37
[62] A. Hayatsu, "K-Ar Isochron Age of the North Mountain Basalt, Nova Scotia," *Canadian Journal of Earth Sciences,* Volume 16, 1976, p. 974
[63] J.F. Evernden, and John R. Richards, "Potassium-Argon Ages in Eastern Australia," *Journal of the Geological Society of Australia,* Volume 9/1, 1962, p. 3
[64] William D. Stansfield, *The Science of Evolution,* Macmillian, 1977, p. 82 & 84
[65] John A. Eddy (as reported by R.G. Kazman), "It's About Time: 4.5 Billion Years," *Geotimes,* Volume 23, 9/78, p. 18
[66] Adrian E. Scheidegger, *Principles of Geodynamics,* 2nd ed. Springer-Verlag, 1963, p. 61
[67] John Reader, *Missing Links,* Little, Brown and Company, 1981, p. 15
[68] *Encyclopedia Britannica,* p. 170
[69] E.M. Spieker, "Mountain-Building Chronology and the Nature of the Geologic Time-Scale," *Bulletin, American Association of Petroleum Geologists,* Volume 40, 8/56, p. 1806
[70] C.F.A. Pantin, "The Origin of Species," *A Short History of Science,* Doubleday, 1951, p. 97
[71] Carl O. Dunbar, *American Journal of Science,* in Historical Geology, John Wiley and Sons, Inc., 1949, p. 52
[72] W.M. Elasser, "Origin, Composition and Age of the Earth," *Encyclopedia Britannica,* 7(1973):850.
[73] L. Harrison Matthews, FRS, Introduction to Darwin's *The Origin of Species*, J.M. Dent & Sons Ltd. London, 1971, p. xi

[74] Richard Fortey, *Fossils, the Key to the Past,* Van Nostrand Reinhold Company, 1982, pp. 28-29, 31
[75] Adolph Knopf, "Measuring Geologic Time," in *Scientific Monthly*, 85(1957):227
[76] Carl O. Dunbar, *American Journal of Science in Historical Geology*, John Wiley & Sons, Inc., 1949, p. 52
[77] William D. Stansfield, *The Science of Evolution,* Macmillan, 1977, p. 80
[78] George McCready Price, *Problems and Methods in Geology*, Christian Evidence League, 1956, p. 6
[79] J.E. O'Rourke, "Pragmatism Versus Materialism in Stratigraphy," *American Journal of Science,* volume 276, 1/76, p. 47
[80] R.H. Rastall, *Encyclopedia Britannica.* 1956, volume 10, p. 168
[81] Eldredge, Niles. Time Frames: *The Rethinking of Darwinian Evolution and the Theory of Punctuated Equilibria.* New York: Simon and Schuster, 1985, p. 52
[82] H.D. Hedberg, "The Stratigraphic Panorama," *Bulletin of the Geological Society of America,* volume 72, 4/61, p. 499
[83] O'Rourke, J. E. "Pragmatism Versus Materialism in Stratigraphy." *American Journal of Science.* (volume 276, 1/76), p. 53
[84] Evernden, J. F., D. E. Savage, G. H. Curtis and G. T. James. "K/A Dates and the Cenozoic Mammalian Chronology of North America." *American Journal of Science,* (Vol. 262, 2/64) p. 166
[85] O.D. von Engeln and K.E. Caster, *Geology,* McGraw-Hill, 1952, p. 423
[86] A.C.M. Laing, "Letters to the Editor." *The Australian Geologist,* Newsletter, no. 48, March 19, 1984, p. 7
[87] Derek Ager, *New Scientist,* November 10, 1982, p. 425
[88] Moshe Pearlman, Digging up the Bible, Littlehampton Book Services Ltd, 1980, p. 54
[89] *Ibid.*
[90] *Ibid.*
[91] *Ibid.*, p. 55
[92] *Ibid.*
[93] H.D. Hedberg: "The Stratigraphic Panorama," *Geological Society of America Bulletin*, 72, April 1961, pp. 499-518
[94] Carl O. Dunbar, *American Journal of Science*, in *Historical Geology*, John Wiley & Sons, Inc., 1949, p. 52
[95] Pearlman, Op Cit.
[96] *Ibid.*, p. 59
[97] *Ibid.,* p. 151
[98] *Ibid.*, p. 64
[99] *Ibid.*, p. 65
[100] *Ibid.*, p. 151
[101] Carl O. Dunbar, *American Journal of Science*, in *Historical Geology*, John Wiley & Sons, Inc. 1949, p. 52
[102] The Foreword to the 1971 edition of Darwin's *Origin of the Species*, Harrison L. Matthews, p. x
[103] David Kits, "Paleontology and Evolutionary Theory," *Evolution*, vol. 28, September 1974, p. 466
[104] In Mayr's book *Systematics and the Origin of Species* (1942), as cited by Niles Eldredge in his book, *Time Frames: The Rethinking of Darwinian Evolution and the Theory of Punctuated Equilibria*, Simon & Schuster, 1985, p. 33
[105] Douglas Futuyma, *Science on Trial*, 1983, p. 152
[106] Colin Patterson, "Cladistics," Interview on BBC, March 4, 1982
[107] Keith S. Thompson, "The Meanings of Evolution," *American Scientist*, vol. 70, September/October 1982, p. 529
[108] H.D. Hedberg: "The Stratigraphic Panorama," *Geological Society of America Bulletin*, 72, April 1961, pp. 499-518
[109] Henry Gee, "In Search of Deep Time," *The Free Press*, 1999, pp. 116-117
[110] Henry Gee, "In Search of Deep Time," *The Free Press*, 2001, p. 23
[111] C.O. Dunbar *Historical Geology*, 2nd ed., John Wiley & Sons, 1960, p. 47

[112] Steven M. Stanley, "Macroevolution and the Fossil Record," *Evolution*, 36, No. 3, 1982, p. 460
[113] Richard Lewontin, "Billions and Billions of Demons," January 9, 1997
[114] H.A. Orr & J.A. Coyne, "The Genetics of Adaptation: A Reassessment," *American Naturalist, 140,* 1992, p. 726

Made in the USA
Monee, IL
08 November 2026

67380771R00083